Introduction to Energy and the Environment

Introduction to Energy and the Environment

Paul Ih-fei Liu

VNR VAN NOSTRAND REINHOLD
New York

Library of Congress Catalog Card Number 93-7906
ISBN 0-442-01557-7

I(T)P   Van Nostrand Reinhold is a division of International Thomson Publishing company. ITP logo is a trademark under license.

Printed in the United States of America

Van Nostrand Reinhold
115 Fifth Avenue
New York, NY 10003

International Thomson Publishing GmbH
Königswinterer Str. 518
5300 Bonn 3
Germany

International Thomson Publishing
Berkshire House 168-173
High Holborn, London WC1V7AA
England

International Thomson Publishing Asia
38 Kim Tian Rd., #0105
Kim Tian Plaza
Singapore 0316

Thomas Nelson Australia
102 Dodds Street
South Melbourne 3205
Victoria, Australia

International Thomson Publishing Japan
Kyowa Building, 3F
2-2-1 Hirakawacho
Chiyada-Ku, Tokyo 102
Japan

Nelson Canada
1120 Birchmount Road
Scarborough, Ontario
M1K 5G4, Canada

16 15 14 13 12 11 10 9 8 7 6 5 4 3 2 1

Library of Congress Cataloging in Publication Data

Liu, Paul I. (Paul Ih-fei)
    Introduction to energy and the environment / Paul I. Liu.
        p.   cm.
    Includes bibliographical references and index.
    ISBN 0-442-01557-7
    1. Energy development—Environmental aspects. 2. Energy consumption—Environmental aspects. 3. Pollution. I. Title.
TD195.E49L58 1993
    333.79'   14—dc20   93-7906
                                                           CIP

To Doktorvater
Prof. Dr. rer. nat. habil. H. St. Stefaniak

CONTENTS

Preface

This book grew out of notes that I prepared for classes held at the University of Southern California and at California State University, Northridge. It was developed and modified from the syllabus used by Drs. T. F. Yen and Joseph Devinny of the University of Southern California.

I have assembled information relevant to the subject from 90 technical sources, presenting environmental problems encountered as a consequence of using energy in industrial practices and describing the technology commonly used to improve environmental quality. The book is an introductory overview of the subject intended primarily to get laypersons, students, and professionals interested in joining task forces in the environmental field and in studying environmental issues.

Special efforts have been made to stimulate the interest of students, in the hope that they may choose a subject area from the topics presented as their major course of study. The book addresses basic concepts of energy and environmental issues so that it can be used as a foundation for professionals. Also, it describes pollution control techniques and waste management strategies so that it can be used as a reference book. The current status of advanced technology relevant to energy and the environment also is cited so that the book can be used as an indicator of future development trends.

The environment encompasses almost everything around us, but it would be unreasonable to include an unlimited array of substances in the limited space of one book. Therefore, this book is concise; descriptions are simple; data are approximate; illustrations and formulae are eliminated whenever possible; and many irrelevant items are omitted. The book provides an extended overview of its subject to government employees and to private consultants in a single volume.

Often during the preparation of this book, I felt that technology was advancing more rapidly than my ability to record it. New knowledge is exploding very rapidly, much faster than one person can grasp. With this challenge in mind, I have gathered information for this book as if it were a basket of water in which I scooped up as much as I could from the ocean, as I sailed on my professional journey, and I have given you a limited sampling of many topics.

Also, I have talked to many other colleagues and friends, including Andrew Lee, Emmanuel Ruivivar, Susan Tsai, Kathy Gee, Marge Valasquez, Dr. Soen Tan, and Wilma LaRocca, as well as discussing the subjects presented here with students in my classes. I am delighted and grateful for their critiques, suggestions, and assistance. My thanks are due to Drs. Christine Liu and Theresia Lee for their review and corrections of the manuscript. I am obliged to Drs. Mihran Agbabian, Joseph Devinny, David S. K. Liu, Charles V. Metzler, Massoud Pirbazari, and Thomas Shen for their encouragement and advice. I owe special thanks to the staff of Van Nostrand Reinhold, especially Judy Brief, Bob Argentieri, Betty Sheehan, and Peter Rocheleau, for without their efforts this book would not be available to you. My love and heartfelt appreciation go to my lovely wife, Johanna, for her patience and support over many years.

I would like to express my thanks to my life-long friend, Carl Schild, a highly respected authority in the fields of electrical and mechanical engineering, who first encouraged me while I was in college. He was often in my thoughts during the preparation of this manuscript.

Summary of Chapters

This book introduces the reader to the relationship of energy, pollution control technology, and the environment. Energy-conserving and environmentally sound advanced technologies also are discussed.

Chapter 1 begins with a discussion of state and federal legislation enacted within the last three decades to protect the environment by controlling air pollution, water pollution, and toxic hazardous waste disposal. To familiarize the reader with frequently used environmental terms, Chapter 1 includes definitions and an explanation of ecosystem structures (atmosphere, hydrosphere, lithosphere), ecosystem components (biotic and abiotic), and ecosystem functions. The relationship of energy to the environment also is discussed.

Chapter 2 starts with a brief definition of renewable resources, and the remainder of the chapter is devoted to discussing renewable resources in the environment and renewable energy. Renewable resources include forests, fisheries, agricultural products, air, and water. The sources of renewable energy are biomass and solar, wind, tidal, hydraulic, and geothermal forms of energy. Important concerns reported in Chapter 2 are the environmental problems generally associated with each source of energy. For example, solar cells tend to be inefficient, wind power generators are not aesthetically pleasing and are noisy, and geothermal energy use can result in odor problems and harmful ammonia waste, which contaminates water.

Chapter 3 discusses natural fossil fuels (coal and petroleum) and synthetic fossil fuels (tar sand, the oil shale retort, and products from coal conversion). Because the focus is the formation of fossil fuels, the chapter explains the carbon cycle as it related to coal formation, as a background for discussing three types of petroleum formations: oil shale, liquid petroleum, and gaseous petroleum. Methods of extracting synthetic fuels (fuels that do not occur naturally) are explained in some detail, along with the environmental concerns accompanying their use.

Chapter 4 is concerned with the problems resulting from burning fossil fuels, these being NO_x emissions and the release of toxic hazardous substances. The two types of NO_x associated with fossil fuel burning, thermal NO_x and fuel NO_x, are explained in formulae showing the atomic reactions of nitrogen and oxygen in several processes. A wide range of NO_x-generating devices, including water heaters, ovens, furnaces, and boilers, is discussed, along with methods to control NO_x formation from these devices. The last section of this chapter is devoted to internal combustion (I.C.) engines, which are major sources of NO_x emissions. All methods commonly used to control NO_x emissions from several types of I.C. engines are discussed, including derating, retarded ignition timing, air-to-fuel ratio adjustment, use of a turbocharge with intercooler, reduced manifold temperature, and others.

Chapter 5 covers five topics: persistency and toxicity of pollutants, environmental impacts and governmental responses, toxic waste disposal technology, developing toxic waste control technology, and sources of toxic pollutants in Southern California. Special attention is given to trace elements such as chromium, arsenic, and lead, which are highly toxic and widespread in the environment. The most desirable methods for controlling toxic substances are discussed, including conversion of toxic hazardous pollutants (incineration, thermal destruction, biological treatment, chemical destruction, land farming, ocean assimilation) and permanent storage (landfill, underground injection, surface impoundment, salt formation, arid region burial). The developing technologies for toxic waste control are discussed, including processes of waste reduction, volume reduction, chemical detoxification, fixation (using chemical additives), and stabilization.

Chapter 6 begins with a brief discussion of how societies historically have handled waste. It explains the types of reprocessing in use and their accompanying air and water pollution problems, and methods of handling waste residues. A major focus is the problems of paper recycling, which involves a pulping process. Combustible refuse recovery (pyrolysis method), ferrous material recycling, and nonferrous material recycling are discussed, including aluminum recycling, copper recycling, and rubber tire resource recovery. Every recycling process produces a waste residue that must be disposed of; therefore, the last part of Chapter 6 is devoted to methods of waste residue disposal and the environmental problems associated with each method.

Chapter 7 is a technical discussion of cooling methods. Most industrial processes are accompanied by waste heat. For environmental and economic reasons, the management of waste heat has become increasingly important. In this chapter the major sources and the dispersal methods of waste heat are discussed, as well as the environmental impacts of several cooling techniques and cogeneration. Cooling methods are classified as once-through (cooling systems that usually are installed in rivers), cooling ponds, and cooling towers. Heat recovery through the use of equipment such as pipes and ducts, air preheaters, recuperators, regenerators, economizers, and so on, also is discussed. All the cooling techniques discussed impact the immediate environment; so this chapter considers these environmental impacts, which usually are adverse.

Chapter 8 presents an overview of oil spills and underground storage tank leakage. It discusses methods of preventing spills and leaks, as well as technologies used to remedy them. The design of a device used for oil slick removal, the oil boom, is described, as well as an oily waste processing plant. The recovery of waste oil sometimes is possible, using modern technologies discussed in this chapter: gravity differential separation, vacuum filtration, acid treatment, electrostatic cleaning, chemical treatment, flocculation and sedimentation, agitation, and ultrasonic vibration.

Chapter 9 begins with an explanation of the chemical composition of petroleum, also known as crude oil, a complex mixture of hydrocarbons and nonhydrocarbon compounds. Many products are manufactured from crude oil, either by rearrangement of the hydrocarbons in the feedstock or by breaking down the complex hydrocarbons into simpler ones, as this chapter explains. There also is a detailed explanation of primary and secondary operations in petroleum refining industries. Primary operations mainly include processes for separation, decomposition, and formation. Within these general categories are subprocess units; for example, when separation processes are discussed, desalting units, distillation units, and deasphalt units are explained. The same approach of explaining processes down to the subprocess unit level is used for secondary operations, which include hydrotreating, gasoline treating, sulfur recovery, and tail gas treatment.

Chapter 10 provides basic principles necessary to an understanding of nuclear power, including an explanation of atoms, protons, and neutrons. A discussion of nuclear power systems and their health effects follows the simplified explanation of nuclear energy. Reaction processes, reactor types, and power systems for fission and fusion energies are covered, and the discussion concludes with a comparison of fission and fusion energy. The characteristics of nuclear radiation are summarized, and their biological effects are explored in terms of damage expected from specified radiation dosages. The final section of this chapter discusses management and disposal technology for radioactive waste, which is categorized as high- and intermediate-level waste. The following waste disposal processes are explored, along with the environmental problems they would present: geological formation on land disposal, an in-situ underground melt process, ocean dumping and subseabed disposal, ice sheet disposal, and extraterrestrial disposal.

Chapter 11 addresses alternative fuels that are under development and have partially proved to be feasible in large-scale applications. These fuels include liquefied petroleum gas (LPG), compressed natural gas (CNG), methanol, ethanol, and hydrogen. The technologies for fuel cells and superconductors also are included. The chapter provides information on the environmental effects of each fuel under assessment. It is important to note that the amount of pollution caused by these fuels and technologies is limited if not negligible. Reaction formulae for methanol, ethanol, the direct hydration process, and hydrogen fuel are shown. Direct and indirect types of fuel cells are reviewed and their electrochemical reactions given. The five types of indirect fuel cells examined are polymer electrolyte, alkaline, phosphoric acid, molten carbonate, and solid oxide fuel cells.

GLOSSARY

Abiotics: Nonliving objects.
ASTM: American Society of Testing Materials.
Attainment Area: A region whose air quality meets National Ambient Air Quality standards.

Biomass: Terrestrial and marine plants that may be combusted directly for heat energy. These include dead trees, leaves, harvest residue, etc.
Biotics: Living objects.
Bituminous coal: Coal having approximately 50 to 80% fixed carbon and 20 to 40% volatile matter, with a low sulfur content and a high heating value.
Breeder reactor system: A nuclear power system that performs a fission reaction and produces a fissionable product in an amount exceeding that consumed in the fission process.

CAA: Clean Air Act (federal legislation enacted in 1963).
CANDU: A nuclear power system that performs a fission reaction using heavy water as the primary coolant and uranium as the fuel.
Criteria pollutants: Those pollutants whose concentrations cause ill effects in humans when they reach a certain level (threshold limit); NO_x, SO_2, PM, CO, O_3, Pb.

Ecosystem: A functional system, including its organisms and physical features.
EGR: Exhaust gas recirculation.
EIR: Environmental Impact Report
Energy: The capacity to do work.
EPA: Environmental Protection Agency.

Flocculants: Fine solids used to accelerate salt removal.
Fuel cells: Electrochemical devices that convert the chemical energy of a fuel oxidant directly into electrical and thermal energy.

Fuel NO$_x$:   NO$_x$ converted from chemically bound nitrogen in fuel.

Geothermal energy: Energy derived from the heat of the earth's interior.

Hydraulic energy: Energy derived from a generator driven by a hydraulic turbine.

Liquefied petroleum gas: A generic expression for propane, butane, or a mixture of both.

Natural fossil fuels: Coal and petroleum obtained from the earth.

Non-attainment area: A region whose air quality does not meet National Ambient Air Quality Standard.

Noncriteria pollutants: Those pollutants that cause adverse health effects regardless of their concentrations, such as asbestos, beryllium, mercury, vinyl chloride, and others.

NO$_x$:   Oxides of nitrogen, primarily NO and NO$_2$.

Oil boom: A device for removing oil slicks.

RCRA: Resource Conservation and Recovery Act.

Renewable resources: Those resources that, after being used, can be brought back to their original state without human effort. Examples include forests and fisheries.

Resource: A natural source of wealth or revenue that can be used to support life and to supply the needs of an organism.

SARA: Superfund Amendment and Reauthorization Act (federal legislation enacted in 1986).

Solar energy: Energy from the sun.

Synthetic fossil fuels (or synfuels):   Those fossil fuels that do not occur naturally.

Thermal NO$_x$:   The thermal fixation of nitrogen in combustion air.

Thermal reaction system: A nuclear power system that performs a fission reaction caused by neutrons.

Tidal energy: Energy exerted to raise the ocean's elevation.

TSCA: Toxic Substances Control Act (federal legislation).

Wind energy: Energy from the wind. It is a portion of solar energy. Unevenly distributed sunlight energy causes the movement of air.

Introduction to Energy and the Environment

1

Overview

From the earliest times when human beings lived in hunter/gatherer communities to the modern era, humankind has adapted and survived. Life continues from generation to generation, and we hope that our offspring will have better opportunities and a better quality of life in a more prosperous social system than our own. Prosperity is closely related to energy, which plays a vastly important role in our daily lives. The use of energy, however, will generate by-products that, unchecked, can devastate our ecosystem. The areas of our concern are air pollution, water pollution, and toxic hazardous waste disposal.

In response to air pollution problems, the U.S. Congress adopted the Clean Air Act (CAA) in 1963, authorizing federal research funds for air pollution related research activities. The CAA underwent several amendments. In 1965, the CAA enacted the establishment of automobile emission standards; in 1970, the Environmental Protection Agency (EPA) was formed, in part to determine National Ambient Air Quality Standards; in 1977, requirements to offset emissions from new sources in non-attainment areas were mandated; in 1990, toxic substances control and global warming research were addressed.

To address water pollution problems, Congress adopted the Water Pollution Act in 1948, the Water Quality Act in 1965, the Water Pollution Control Act in 1972, and the Safe Drinking Water Act in 1974. In these acts, the legislative control of water pollution evolved from its beginnings as a general statement to become an implementation plan, and later to legislation requiring the installation of the best available water treatment equipment and the determination of maximum concentration levels for specific chemical compounds in drinking water.

As a supplement to these responses, Congress adopted the Resource Conservation and Recovery Act (RCRA) in 1976 and provided EPA with the authority

1

to regulate toxic waste disposal on land. In 1976, the Toxic Substances Control Act (TSCA) was adopted to protect the populace from injury due to exposure to chemical substances. In 1980, the Comprehensive Environmental Response, Compensation, and Liability Act (CERCRA) was adopted, known as the Superfund Act, which requires private parties to accept responsibility for their release of toxic hazardous waste. In 1984, RCRA was amended to ban landfill for untreated hazardous waste, and in 1986, the Superfund Amendment and Reauthorization Act (SARA) required responsible parties to take corrective measures, remediating damages that had occurred.

In California, the control measures for toxic hazardous materials were taken one step further: in 1983, Assembly Bill 1807 was introduced by Assembly-woman Sally Tanner, and became known as the Tanner Bill. The bill mandated that toxic air pollutants be controlled to levels that prevent harm to the public health. The state is required to identify toxic air contaminants and to adopt the appropriate control measures. Local air pollution districts must incorporate these control measures into their rules and regulations. In 1987, California voters enacted Proposition 65, which prohibited the release of any carcinogen or mutagen into any source of drinking water, and required that warnings be posted to alert persons to the possibility of exposure to carcinogenic and/or mutagenic substances. In 1987, Assembly Bill 2588, Air Toxics "Hot Spots" Information and Assessment, was adopted to require the reporting of toxic uses by industry.

This book covers the topics discussed by persons interested in energy and the environment. It includes discussions of conventional energy sources, their effects on the environment, and new energy sources. The writer hopes that the materials included here will provide a background for the discipline of environmental engineering. The basic concepts of ecosystem, energy, and the environment are summarized herewith.

ECOSYSTEM

This section includes the definition, structure, components, and function of an ecosystem.

Definition of Ecosystem

An ecosystem is a functional group of interdependent parts, including organisms and physical features. An organism is any living entity, either animal or plant, having parts or organs that work together as a whole to maintain life and its activities. Physical features would normally include classifications such as desert, mountain, and forest. In an extended sense, physical features also include urban, suburban, and farmland settings.

Structure of an Ecosystem

The structure of an ecosystem can be divided into three components: atmosphere (air), hydrosphere (water), and lithosphere (earth).

Atmosphere
Two terms describe the constantly changing conditions of the atmosphere: climate and weather. Climate denotes an average of atmospheric conditions over a long period of time (say ten years). Weather denotes the day-to-day atmospheric conditions. The movement of air carrying released exhaust gases and toxic chemical compounds results in air pollution problems, which are our concern in Chapters 4 and 5.

Hydrosphere
The hydrosphere component includes streams, rivers, lakes, oceans, and glaciers. Water pollution problems, especially those related to waste heat management and oil spills, are closely associated with the hydrosphere (Chapters 7 and 8).

Lithosphere
The lithosphere is the soil, which consists of three layers.

1. The top layer, or A-horizon, contains plant leaves, other organic fragments, and inert dust. This layer retains nutrients and supports plant life.
2. The subsoil layer, or B-horizon, receives inorganic materials, such as calcium, aluminum, and iron, from the A-horizon to support human life.
3. The rock layer, or C-horizon, supports the soil structure. Proper solid waste management, such as recycling of solid waste (Chapter 6), and the control of underground storage tank leakage (Chapter 8) are vital to the lithosphere.

Components of an Ecosystem

The components of an ecosystem can be categorized into two groups: biotic and abiotic. Biotics are living objects; abiotics are nonliving objects.

Biotic groups are made up of producers and consumers. Producers are plants, vegetables, and those entities that produce food from water, carbon dioxide, and sunlight to complete the photosynthesis process. Consumers are the users of this food or of other organisms to obtain nutrition and energy for their own survival. Consumers are subdivided into four classes:

1. Herbivores, which eat plants only.
2. Carnivores, which eat meat only.

3. Omnivores, which eat both plants and meat.
4. Detritus consumers, such as bacteria, earthworms, oysters, and decomposers, which do not eat plants or animals although they breathe air.

Abiotic entities are chemical substances, which include inorganic and organic chemical compounds. Inorganic chemical compounds are substances other than animal or plant matter. Organic chemical compounds are fats, proteins, vitamins, and other substances containing carbon atoms—except for carbon monoxide, carbon dioxide, carbonic acid, metallic carbide, and metallic carbonate.

Function of an Ecosystem

The function of an ecosystem is revealed in the movement of matter and energy within the system. Based on the particular organisms present, all activities in an ecosystem are due to the energy produced by the movement of matter in the form of a food chain.

Matter and energy occupy an ecosystem whose ultimate source of energy is the sun. Energy transmitted from the sun supports and maintains life on the earth.

ENERGY

Definition and Properties

Energy is the capacity to do work. Its properties are described by the first and second laws of thermodynamics: The first law states that energy can neither be created nor destroyed; it can simply be converted from one form to another. The second law states that a conversion of energy always produces some less useful form of energy, usually heat energy.

Forms of Energy

Energy exists in various forms, depending on the energy source. The forms are summarized below:

- Heat energy: due to random motion of particles.
- Mechanical energy: due to speed (dynamic energy), elevation (potential energy), or movement (kinetic energy) of an object.
- Electrical energy: due to movement of charged particles.
- Chemical energy: due to the energy contained in chemical bonds.
- Nuclear energy: due to binding energy of atomic nuclei.
- Gravitational energy: due to gravitational attraction.
- Light energy: due to electromagnetic radiation.

Energy Resources

Energy can be obtained from renewable and nonrenewable resources. Renewable resources are discussed in Chapter 2. Nonrenewable resources include fossil fuels and nuclear energy. Fossil fuels are formed via the preservation of ancient organisms under special conditions (Chapter 3). As a result of the combustion of fossil fuels (Chapter 4), energy is released, which can be converted into other types of energy. Nuclear energy—fundamentals and nuclear power systems—is described in Chapter 10.

THE ENVIRONMENT

Energy exists in different forms; harnessed energy enhances our well being, but the consumption of energy creates by-products that can damage or destroy our surroundings or the environment. These by-products can be in particulate, gaseous, or wave form. They appear as toxic waste, waste heat, or radioactive materials. Means of controlling particulate matter and gaseous emissions are summarized in Appendixes A and B. Toxic waste management and its technology are described in Chapter 5. The recycling of solid waste including emission control strategies and the management of waste heat are discussed in Chapters 6 and 7. Chapters 8 and 9 discuss oil spills, underground storage tank leakage, and petroleum refining operations. The health effects of nuclear power and radioactive waste management are presented in Chapter 10. To highlight the future development of energy-conscious and environmentally sound technologies, alternative fuels, fuel cells, and superconductors are discussed in Chapter 11.

2

Renewable Natural Resources

This chapter defines renewable resources and discusses renewable resources in the environment and renewable energy sources.

DEFINITION OF RENEWABLE RESOURCES

A resource is a natural source of wealth or revenue that can be used to support life and to supply the needs of an organism. Renewable resources are those resources that, after being used, can be brought back to the original state without human effort. Proper management of resources is required to maintain their quality; otherwise the ecosystem will lose its balance, and society will be adversely affected.

RENEWABLE RESOURCES IN THE ENVIRONMENT

Renewable resources include forests, fisheries, agricultural products, air, and water.

Forests

A forest is considered a renewable natural resource because the seeds or branches of its trees are expected to replace the portion of the forest used for either construction materials or energy sources. Well-planned, properly managed forest use is necessary to maintain forest quality. Thinning of overcrowded forests to provide better growing conditions and to increase the sources of wood for fuel and lumber is a major task in forest management.

6

Fisheries

Fisheries are considered a renewable natural resource because fishing occurs in a natural environment unmodified by humans. Approximately 70 percent of the surface area of the earth is covered by water; there are enough fish (from trawling tuna fishing, etc.) in the ocean to supply the protein needed by human beings.

Normally, fish are concentrated in cold areas, north of 40° latitude or south of –40° latitude, for the following reason: Nutrients usually float on the upper layer of the ocean, but in the wintertime the water in the lower layer of the ocean is warmer than that in the upper layer, so that nutrients move from the upper layer downward. Thus the nutrients are distributed to best supply the fish.

Steps undertaken in practice to manage fish populations accomplish the following:

1. Make possible cultivation of fish in hatcheries, artificial ponds, or tanks.
2. Remove competing fish from the area.
3. Put limits on the time spent in fishing.
4. Set an allowable quantity of fish that may be kept.
5. Restrict the type of gear used.

These activities regulate the amount of fish caught to ensure that adequate numbers of fish remain for breeding.

Agricultural Products

Agricultural products include vegetables, fruits, and crops, which are the result of the fixation of solar energy by green plants to make products that will be consumed by humans and animals.

The growth and the productivity of green plants depend on soil and climate. Soil is a mixture of inorganic material (minerals, rocks, water, and air), organic matter (animal droppings, plants, and other hydrocarbons), and living organisms (bacteria, worms, and insects); it is the support medium for the root systems of green plants and the depository of nutrients and water. Climate determines the distribution of light, heat, and rain, thus regulating cycles of plant growth and levels of productivity.

Agricultural products are classified as a renewable resource, for which proper management is required to determine:

• When fertilizers should be released.
• How weeds are to be controlled.
• Which insects and diseases must be removed.
• When fossil fuel is to be used.

The quality of soil and the surrounding temperature determine the overall yield of agricultural products.

Air

In its formative period, the entire earth was covered with methane and ammonia gas. Later, owing to cosmogonic effects and volcanic explosion, the composition of air was changed to include approximately 78 percent nitrogen, 21 percent oxygen, and 1 percent in total of argon, carbon dioxide, water, ozone, aerosols, and air pollutants. Although air is classified as a renewable natural resource, and the consumed oxygen in air is refurnished through photosynthesis, air pollutants must be controlled to maintain the quality of air.

Air pollutants are classified into criteria and noncriteria pollutants. Criteria pollutants are those whose concentrations cause ill effects in humans when they reach a certain level (threshold limit). They include nitrogen oxides (NO_x), sulfur dioxide (SO_2), carbon monoxide, ozone, particulate matter, and lead. Noncriteria pollutants are those that cause adverse health effects regardless of their concentrations. They include asbestos, beryllium, mercury, vinyl chloride, and other toxic hazardous materials.

These air pollutants can be controlled by one of the following devices or methods.

Particulate Matter Control (Appendix A)

Electrostatic Precipitator
An electrostatic precipitator collects particulate matter by charging it through a high voltage electrode wire. When a dust-laden stream passes the collector, negatively charged dust particles are accelerated toward collecting plates that are grounded (positively charged). Through plate rapping, the accumulated dust particles are released from the plate and dropped to a collection hopper, where the dust is transported through a screw-conveyor and removed from the system.

Baghouse
A baghouse is a bag filter. When dirty air approaches a filter, particulate matter is intercepted, and only clean air passes through the filter. The filter then is shaken or blown with a stream of reverse-jet air to release the accumulated dust from the bag. The dust falls to the collection hopper for removal.

Scrubber
A scrubber is a wet-type collector. Liquid droplets are sprayed onto dust-loaded gases to wet the particles that impinge on collecting surfaces. The collected dust settles to the bottom of the sump to be removed from the system.

Cyclone
A cyclone is a centrifugal separator. Dust-loaded gases enter a cylindrical tube, usually tangentially, causing centrifugal force to drive particles out of the main gas stream. The coarse particles are thrown out, impact the cyclone's inner wall, and drop down along the wall for removal.

Gaseous Emission Control (Appendix B)

Complete Combustion
Combustion is the most effective means of destroying gaseous emissions. Complete combustion can convert organic compounds into harmless carbon dioxide and water by rapid oxidation. Special combustion technology tactics, such as turbulent mixing, adjustment of the air-to-fuel ratio, and control of the time of exposure to peak temperature, should be incorporated into practice (Chapter 4).

Gas Absorption
Gas absorption is an operation in which soluble components of a gas mixture are dissolved in a liquid. They dissolve primarily because of liquid–gas interaction, the amount of gas dissolved in the liquid depending upon the type of chemical compound and its concentration in the gas stream. The dissolved chemicals usually are recovered.

Adsorption
Adsorption is the capture of gas components and their adhesion to the surfaces of solid bodies with which they are in contact. It is done in a solid–gas contact device that mainly employs activated carbon. Once the solid surface is saturated with the gas compound, the system will turn from the adsorption mode into a regeneration process; steam usually is injected to strip the adsorbed compounds from the carbon, and the effluent is sent to the recovery system. The regenerated carbon is returned to the adsorption device to start the adsorption process anew.

Condensers
Condensers convert vapors into liquid form. When the heat energy is removed, heated vapors cool and condense to form a liquid, which can be disposed of in deep wells or treated separately.

Other Methods
Gaseous emissions can be controlled by using scrubbers to remove impurities. (See Appendix B.) By changing or controlling a process, modifying equipment, or applying properly selected process materials, gaseous emissions can be reduced.

Toxic Hazardous Emission Control
Details of this type of control are discussed in Chapter 5.

Water

Water can be converted from one phase to another: It can be evaporated from a liquid to a gaseous form; it can be condensed from a gaseous form to liquid droplets. Water molecules from glaciers are sublimated from solid ice into the air; and water can be deposited as a solid by being changed from a vapor form to hail. With all the conversion of water that occurs from one phase to another, the total amount of water around the earth remains almost constant.

Water is a renewable natural resource whose quality can be maintained only by proper control of pollution. Water pollution occurs in both surface water and groundwater. Surface-water pollutants are infectious agents, oxygen-demanding wastes, cultural eutrophicants, toxic substances, spilled oil, waste heat, and sediments. Groundwater pollutants are calcium, magnesium, pathogens, and nitrate. The content of these pollutants must be reduced substantially to meet the water quality standards specified in the Safe Drinking Water Act.

RENEWABLE ENERGY

The sources of renewable energy are biomass and solar, wind, tidal, hydraulic, and geothermal forms of energy. The recovery of renewable energy generally is subject to a number of problems:

- A large parcel of land is required.
- A large amount of material is required for facility construction.
- Health and safety problems develop during the facility's operation.
- A large amount of energy consumption is needed for the collection of renewable energy.
- Air pollution and water pollution by chemicals result from processes related to the recovery of renewable energy.

Some specific problems related to each source of renewable energy are discussed briefly in the following paragraphs.

Biomass

Biomass, which includes dead trees, leaves, harvest residues, waste wood, and sawdust from wood or carpentry, can be used to produce energy. Specific problems related to the use of biomass for energy are the following:

- Biomass requires a large parcel of land to reproduce the source.
- Biomass collection requires energy.

- The process of producing energy from biomass creates air pollution problems.
- The process also creates health and safety problems.

Yunnan Province, in Xuan Wei County in China, has had the highest rate of lung cancer in China for both its male and its female populations. However, the percentage of female smokers is far lower than that of male smokers; thus smoking is not the sole cause of lung cancer there. In 1982, through a cultural exchange program, EPA scientists discovered that polycyclic organic compounds were being released from coal and wood-burning stoves in Yunnan Province. Polycyclic organic compounds, which are carcinogens (Chapter 5), are by-products of incomplete burning of the biomass grown in that area; and the biomass is the main source of energy used for cooking purposes there.

Solar Energy

Solar energy, the energy from the sun, gives life to the earth. The solar power that falls upon the earth is estimated as 180 trillion (180×10^{12}) kW, which is nearly 500,000 times the electric power capacity of the United States.

Solar energy has been used in various devices, from small, simple heating systems to ground-based thermal plants, photovoltaic systems, and space satellite applications. Small, simple heating systems include garden hose, pipelines, and water containers exposed to sunlight.

Ground-based thermal plants use a piping system enclosed in boxes that have glass on the top and a black metal plate at the bottom. The greenhouse effect enables the boxes to become solar energy collectors that heat the incoming cold water. The heated water is accumulated in a water tank and is pumped through an upper section of the tank to a heating coil to warm the facility. After the release of the heat, cold water is returned to the lower section of the water tank and then is pumped to the piping system in the collector. These plants can be used only during the daytime when sunlight is available.

A photovoltaic system converts sunlight to electricity directly. Solar cells use semiconductor materials, usually silicon, that when struck by solar radiation give rise to a current. When two silicon thin plates (p and n types) are put together, an inherent electric field is formed. When sunlight penetrates the surface of the cell, electrons can be excited to higher energy levels by interaction with photons (elements that have no mass, no electric charge, but only motion in a wave form at the speed of light). This energy transfer, between photons and electrons, allows electrons to migrate across the barrier. The migrating electrons are guided through a current collector, thus becoming the electrical power that is generated.

Solar cells are quiet, reliable, and easy to operate, and they have no moving parts. In space satellites and orbiting space stations, these cells are a useful power source. The efficiency of solar cells is estimated as 15 to 40 percent, depending upon the materials used.

Wind Energy

Wind energy is a kind of solar energy; unevenly distributed sunlight energy causes the movement of air. The kinetic energy of the wind thus produced is converted to electric energy.

Wind power is equal to the kinetic energy per unit time. It is proportional to the product of the air mass times the square of the wind velocity, or to the cube of the wind velocity. Therefore, a small increase in wind velocity will cause wind power to increase tremendously.

Wind power energy is estimated to be one trillion (1×10^{12}) kW worldwide. The efficiency of the net energy conversion has been estimated as 12 percent.

A windmill usually is used to convert wind energy to electricity. The structure of a windmill consists of blades mounted on a high tower. An air shaft transmits the blades' rotation to a gear box, which converts the shaft rotation from approximately 40 rpm to 1800 rpm power to drive a power generator. The power that is generated is stored in batteries for future use.

Wind power generation mainly affects the environment by being aesthetically unpleasant. Also, noise pollution, falling blades, and structural defects due to vibration can cause injury and damage. Wind power generators divert large areas from regular land use. They also interfere with TV and radio reception, and they change the flight paths of birds, thus affecting the birds' migratory habits. All these are considered negative environmental effects.

Tidal Energy

Tidal energy is defined as the energy used to raise an ocean's elevation. This form of energy is mainly due to the earth's rotation and gravitational effects from the sun and the moon on the earth.

The tidal effect on a particular location can be determined from long-term statistical records for that location. Predictions of when and how the sea level falls are particularly important for harbors where shipping is dependent upon tidal levels. Tidal energy can be utilized to lift ships in an outer harbor to an elevation comparable to the sea level of the inner harbor water basin. Then the ships can travel in or out of the harbor.

Tidal energy can be converted into electrical power energy. The capacity of tidal power has been estimated to be 36 trillion kWh per year worldwide. Near the Ranch river in Normandy, France, a tidal power plant has been in operation since 1966. It consists of twenty-four 10 MW (1 watt = 746 ft-lb) turbine units. The plant operates on 40-foot tides and produces 500×10^6 kWh annually. Tidal falls are guided to move turbine blades to generate power at the peak hour. The spent seawater is pumped to a reservoir during the off-peak hours when the turbines are run as pumps.

The Passamaquoddy River, which flows between the state of Maine and Canada, has a potential of 1800 MW and offers the only important prospects for tidal power in the United States. However, a power plant was not built there because its estimated electrical output was not considered economically competitive.

Hydraulic Energy

Energy radiates from the sun into space, and some reaches the earth. There it warms the atmosphere and causes water to evaporate from oceans, lakes, and rivers. Along with some dust and aerosols, the evaporated moisture condenses and precipitates as rain. The rainwater can be accumulated and stored in a dam at a high elevation. To produce energy, water is released from the dam to convert potential energy into kinetic energy and then into electric energy.

To generate hydraulic energy, water is directed to fall on a hydraulic turbine that turns a shaft coupled to an electric generator, which produces voltage and supplies electric energy to a load. The overall efficiency of the energy conversion is approximately 90 percent. Hydraulic power plants also can be used for flood control, irrigation, recreation, navigation, and public water supplies.

The production of hydraulic power influences the environment in several ways:

- The areas used for hydraulic reservoirs may have more desirable uses than providing electricity.
- Fish migration is restricted by these reservoirs.
- The health of fish populations is affected by changes in the reservoir water temperature.
- Dam construction is costly.
- Safety provisions are needed for areas downstream of the dam.

Geothermal Energy

There are approximately ten crustal plates on the earth. Some plates carry mountains and oceans, some carry deserts, and others carry other geologic features. The plates move relative to each other, grinding and overlapping, and in different directions. The areas near crustal lines, such as Japan, Turkey, the western side of South America, and the San Andreas Fault in California, have earthquakes and other volcanic activities.

Underneath a soil layer, groundwater flows on a porous rock layer on top of a layer of solid rock. Below the solid rock layer, hot magma, which is molten rock, generates heat that is transferred to the water that has seeped down in the porous rock layer. When the water that has seeped into the porous layer is boiled, steam is released through crustal lines and appears on the surface as fumaroles or geysers.

The steam delivered from a crustal line through a well carries geothermal energy, as well as hot molten materials that must be separated from the vapor by special treatment. The treated steam is guided to a steam turbine coupled to an electric power generator.

The advantages of using geothermal energy are the following:

- No irreplaceable fuel is used.
- Only a small land area is needed.

The use of geothermal energy also has negative environmental effects:

- The escaping steam creates noise pollution.
- Odor problems result from sulfur and ammonia release.
- There are aesthetic concerns (poor scenic views due to equipment installation, for example).
- The wastewater contains ammonia, which harms fish and plants.
- There is danger of land subsidence.

Along with the renewable energy sources discussed above, fossil fuels have long been an important energy source. Their formation is described in the following chapter.

3

Formation of Fossil Fuels

This chapter discusses natural and synthetic fossil fuels. For natural fossil fuels, the formation of coal and petroleum, including some of their properties, is discussed. For synthetic fossil fuels, the tar sand process, the oil shale retort, coal conversion, and environmental concerns are presented.

NATURAL FOSSIL FUELS

The word "fossil" comes from a Latin word meaning "dug up," the implication being that fossil fuels are substances that can be obtained by digging into the earth, or are those that are the leftovers or the residuals of ancient ecosystems that have been saved from complete oxidation.

Fossil fuels consist of coal and petroleum, and they include crude oil, natural gas, and oil shale. The main sources of fossil fuels are terrestrial trees and marine organic matter.

Formation of Coal

A carbon cycle occurs in nature: Carbon dioxide in the atmosphere and water vapor react in the presence of sunlight to produce protein and oxygen, in a process known as photosynthesis. Protein appears in green plants and vegetables, and is consumed in the plants' respiration process; carbon dioxide is released to the atmosphere to complete the carbon cycle. More protein is fixed in the photosynthesis process than is used in the respiration process of green plants; the balance is accounted for by animal consumption and bacterial decomposition of dead trees.

Sometimes, because of earthquakes or other accidents where land falls, terrestrial trees are trapped and buried by soil or rocks. Leaves and branches may fall into such traps and accumulate to form a layer of organic matter. This organic matter containes an abundance of woody wall cells (known as lignin) and a fibrous substance (cellulose). With time, the temperature and the prèssure increase in the area surrounding the organic matter, and water and organic gases are driven off. The compressed organic matter has a high carbon and a low water content. Finally, mineral matter is formed, known as coal.

Coal does not migrate and is found in thick continuous layers between rock zones. It is composed of carbon, hydrogen, sulfur, nitrogen, and oxygen.

Formation of Petroleum

Petroleum is defined as an oily liquid solution of hydrocarbons that occurs naturally in rock. It includes oil shale, liquid petroleum, and gaseous petroleum.

Oil Shale Formation

In shallow marine environments, coastal areas, or aquatic basins, aquatic organic matter is trapped when there is a rise in sea level or flooding in the coastal area. These materials are deposited in fine-grained sediments that gradually are compacted and converted into rock form. This aquatic organic matter contains a large amount of fat (lipids), protein, and carbohydrates.

When the organic matter is deposited where there is no oxygen (say up to 1000 m deep), so-called anaerobic decomposition occurs. Organic matter is dissolved in the surrounding moisture, where microorganisms convert it to alcohol and finally produce methane. The residues of the anaerobic decomposition are oily insoluble organic compounds, known as kerogen, a precursor to petroleum. The process of kerogen formation is known as diagenesis.

Rock containing low concentrations of kerogen is called source rock, and that containing higher kerogen concentrations, which can be cost-effectively refined, is oil shale. The chemical elements in kerogen are carbon, sulfur, oxygen, nitrogen, and hydrogen.

Liquid Petroleum Formation

The buried organic material becomes more compact with increasing depth, and its porosity and permeability decrease. Because of temperature increases (to approximately 100°C) at further depth (say up to 2000 m), diagenesis ceases, and thermal reaction becomes important; water vapor, if any, is removed, and chemical bonds are broken. Sulfur, oxygen, and nitrogen are driven off, and kerogen begins to decompose into more mobile molecules that may migrate out from the source rock and flow into more porous and permeable places where oil accumulation occurs. This material is the source of crude oil, and the process of converting kerogen into petroleum by means of thermal reaction is known as catagenesis. The chemical elements in liquid petroleum are carbon, hydrogen, sulfur, and nitrogen.

Gaseous Petroleum Formation

At further depth (say up to 3000 m), where the temperature increases to approximately 150°C and pressure rises to several hundred psi (say 500 psi), the carbon bonds are broken, resulting in lighter molecular weight gaseous hydro-carbons known as gaseous petroleum, which is the main source of natural gas. The gaseous petroleum migrates upward, or from a high-pressure to a low-pressure zone. The major elements of gaseous petroleum are carbon, hydrogen, and nitrogen.

Deeper in the earth is the graphite layer, a soft black natural form of carbon. Graphite has been used for electrodes, lubricants, and pencil lead. An extremely volatile substance, it also is used in the control of nuclear energy.

Some Properties of Fossil Fuels

Natural Gas

The composition of natural gas is approximately the following:

Nitrogen, N_2	5%
Methane, CH_4	80%
Ethane, C_2H_6	10%
Propane, C_3H_8	4%
Butane, C_4H_{10}	1%
(and others)	

Natural gas is a clean fuel whose combustion products are mainly carbon dioxide and water. It should be used for domestic purposes, or when air quality is of vital concern.

Crude Oil

Crude petroleum is poured into a fractionator that distills and cracks the crude into six fuel oil grades based on ASTM (American Society for Testing and Materials) specifications:

- No. 1, a straight-run distillate, used almost exclusively for domestic heating (slightly heavier than kerosene: rocket and jet engine fuel).
- No. 2, a straight-run or light cracked (Chapter 9) distillate used as a general-purpose domestic or commercial fuel in atomizing-type burners.

- No. 3, a slightly heavier cracked distillate which is not commercially available.
- No. 4, a heavy straight-run or cracked distillate that is used in commercial or industrial burner installations.
- No. 5, a lighter-residuum fuel oil that is burned, usually under weather-related conditions, without preheating (in locomotives, ships).
- No. 6, a heavy-residuum fuel oil that usually must be preheated before being burned.

Fuel oils are easily transported and are used mainly in transportation vehicles, to fuel power plants, and for agricultural purposes.

Coal

Coal can be divided into two major categories: bituminous and subbituminous. Bituminous coal is produced in the eastern United States, and has a high heating value, but also is high in sulfur. Subbituminous coal is obtained in the western United States. It has a low heating value and a low sulfur content, but has a high ash content.

In places where there is a shortage of fuel oil, and nuclear power is banned because of its risks, the use of coal-fired power plants can be considered as a major source of electrical energy. Elsewhere, where the technology is advanced, coal is converted to liquid or gaseous form for use in the manufacture of synthetic fossil fuels.

SYNTHETIC FOSSIL FUELS

Synthetic fossil fuels, or synfuels, are those fossil fuels that do not occur naturally. Methods for making synfuels are discussed in the following paragraphs, with environmental concerns cited at the end of the chapter.

Tar Sand Process

Tar sands are mixtures of sand grains, water, and bitumen, which is a viscous, dense petroleum substance that adheres to the sand, giving a black coloration to the tar sand and the surrounding area. Tar sands have a layered structure. Water layers encompass sand particles, forming water–tar sand cells. Bitumen adheres to groups of cells, wrapping them in a film, to produce tar sand units that spread on a tar sand bed.

Tar sand beds exist in rivers such as the Athabasca River in Alberta, Canada. The process of obtaining synthetic crude from the tar sands includes mining, extraction, separation, recovery, coking, and hydrotreating (Chapter 9).

Tar sands are mined under cold and noisy conditions. Then they are guided into a rotating hot-water drum where they are spun until the mixture becomes a slurry-type pulp. This is the primary extraction step.

Next steam is injected, and the pulp is screened. Because of its high sulfur content, the mixture is kept at a pH of 8 to 8.5. The resulting liquid is pumped into a separation chamber, where naphtha is added and air bubbles are introduced into the bottom of the chamber. The bubbles move upward to separate the bitumen from the sand.

Naphtha is added to recover the water (dehydrate), to dilute the bitumen (demineralize), and to separate the water from the sand. The naphtha is recovered in a solvent recovery system, and the liquid is transferred to a coking system that uses high pressure and heat in the absence of air to drive off hydrogen sulfide (H_2S).

To obtain a better-quality bitumen, further hydrotreating is necessary. This is accomplished by adding pressure (up to 1000 psig) and using different catalysts to remove sulfur, nitrogen, oxygen, and halogens (I, Br, Cl, F, At). The product thus obtained is a high-quality synthetic crude.

Oil Shale Retort

Oil shales are mined by using mechanical and electrical equipment or chemical explosives. The shales then are passed through other equipment for crushing, retorting, separation, and hydrogenation (Chapter 9) processes.

Underground mining is noisy, and it produces exhaust gas containing free silicon. The chemical explosives used include TNT (trinitrotoluene) or picryl chloride, an explosive substance with an accompanying fire hazard.

After the mining process, the oil shales are transferred to a crusher for pulverization. Later the pulverized raw shales are conveyed and dumped into an opening at the top of a retort.

A retort can be divided into several zones. At the top, there is an opening into which the raw shales are fed; and below the raw shales, there are zones for mist formation, retorting, heating, dropping, and cooling. In the heating zone, oil, coal, or gas is burned to supply heat. The shales are heated, decomposed, and distilled in the retorting zone. In the mist-formation zone, oil vapor and shale gas are formed. The residues are dropped in the dropping zone, and are cooled before the spent shales are removed.

The oil vapor and the shale gas are guided from the mist-formation zone to the separator, where oil and gas are separated. The oil is accumulated in a storage tank while the gas and the nonseparable oil mists are guided to an electrostatic precipitator. The oil mists are collected by the electrostatic precipitator and guided to the storage tank. The gas is vented by a recycling blower; it can be recycled back to the retort, or it can be used as the heating medium in a heat exchanger, as the coolant in a cooling system, or as a gas product.

The shale oil collected in the storage tank can be passed through a hydrotreating process to upgrade it.

Coal Conversion

Coal can be converted to a gaseous fuel by coal gasification, or converted to a liquid fuel by coal liquefaction.

Coal Gasification

The process of coal gasification includes several steps: mining, crushing, gasifying, separation, quenching and scrubbing, shift-conversion, gas purification, and methanation.

After mining, the coal ores are transported to a storage facility where they are conveyed to primary and secondary crushers. Then the pulverized coal is injected onto the chamber bed of a gasifier while hot steam and oxygen gas are fed in through the bottom of the bed. The coal powders, steam, and oxygen gas are well mixed, creating turbulent whirls in the chamber. As it is homogenized, the mixture possesses the properties of a fluid, characterized by turbulent waves, constant temperature, and the capacity to float around inside the chamber. The system is known as a fluid bed, or a fluidized-bed system.

The oxidized ash is removed from the bottom of the bed while the product gas is vented from the gasifier to a separator. The separator is a typical cyclone where large solid particles are removed by centrifugal and gravitational forces, and the gas with small particles is ducted to a quencher. In the quencher, the small particulate matter is scrubbed; the water-soluble chemicals such as hydrogen sulfides, carbon monoxide, carbon dioxide, and methane are dissolved; and trace elements are partially removed from the gas stream as it is forced through a shower of water.

The semicleaned gas stream is guided to a shift-converter where hot steam is injected into the gas stream. Carbon monoxide and steaming water vapor react in the presence of a catalyst to produce carbon dioxide and hydrogen. A specified hydrogen–carbon monoxide ratio (say 3 to 1) can be obtained by proper control of the catalytic reaction.

When the desired hydrogen–carbon monoxide ratio is reached, the gas stream must be purified to remove sulfur, hydrogen sulfide, carbon dioxide, water, and any remaining hydrocarbons. Various methods of gas purification may be used:

• Hot potassium carbonate solution and diethanolamine (or diglycolamine) may be injected into the gas stream.

- The gas stream may be chilled and washed with cold methyl alcohol.
- The gas stream may be desulfurized.
- Hydrocarbons may be removed from the gas stream.

The purification of the gas stream is necessary to prevent impurities from poisoning the catalyst used in the methanation step, which is described below.

In the methanation step, the gas stream is guided through a nickel catalyst at high pressure (typically 500 psig) and high temperature (say 3000°F) to convert hydrogen and carbon monoxide to methane and water:

$$3H_2 + CO \xrightarrow{\text{(Ni Catalyst)}} CH_4 + H_2O$$

The methane thus produced is the major component of synthetic gaseous fuel.

Coal Liquefaction
Coal liquefaction is the conversion of coal to synthetic low-sulfur, low-ash liquid fuels. The product, syncrude, is a material used for petrochemical products and is suitable for use as a refinery feedstock. A wide range of liquid products can be produced, such as heavy fuels for power plants, distillate fuel oils for commercial uses, and gasoline for transportation. Methods of direct and indirect liquefaction commonly are used.

Direct Liquefaction
Pulverized coal is mixed with solvents (possibly naphtha) and a recycled liquid product, becoming a coal slurry. The slurry is supplied with hydrogen and guided to a reactor where syngas is driven off and highly aromatic liquids containing high levels of sulfur and nitrogen are produced. The process uses a continuous-flow slurry bed and continuous tubular flow. Typical liquefaction conditions in the reactor are 800°F and 1500 psig.

The high sulfur content syngas is ducted to a sulfur recovery unit where the gas stream is preheated, mixed with air, and fed to a catalytic reactor followed by a condenser. The efficiency of sulfur recovery is 80 percent or greater. The desulfurized gas stream is the fuel gas product. The slurry is transferred from the reactor to a fractionator, where it is broken up into different portions by a distillation method (Chapter 9). The products include fuel gas, naphtha, gasoline, heavy oil, and a solid material residue. The heavy oil is recycled back to form a coal slurry.

Indirect Liquefaction
In indirect liquefaction, coal is converted to a methanol solution (with 17% water) by using a coal gasification method, after which the methanol solution is passed

over a catalyst where it is converted into a liquid, a mixture of gasoline and water. The efficiency of the energy conversion is approximately the following:

* For coal liquefaction: 78 percent.
* For coal gasification: 65 percent.

Environmental Concern

Research on carcinogenesis, mutagenesis, and teratogenesis due to exposure to synthetic fossil fuels is being conducted by industrial and governmental agencies. Some results of this research are summarized below.

Tar Sands
* Sand grains of tar sands contain 40 to 50 percent silicon, a known carcinogen.
* During mining operations, both noise and cold stress (at the subzero temperatures on the job sites in Northern Alberta, Canada) are major concerns.
* Naphtha is involved in the processes of extraction, separation, and solvent recovery. The aromatic compounds contained in naphtha, such as benzene, are known to be carcinogenic. Activated carbon adsorbers and thermo-oxidizers have been used as control equipment.
* The gases driven off in the coking process, which contain hydrogen sulfide, carbon monoxide, and polynuclear aromatic hydrocarbons, must be monitored carefully. The last-named compounds are known carcinogens. An amine scrubber and a flare can be used to control the emissions. (Aromatic hydrocarbons and hydrogen sulfide also are emitted from the hydrotreating process.)

Oil Shale
* The shale oil itself is carcinogenic although upgraded or hydrotreated shale oil is essentially noncarcinogenic.
* The mined oil shale contains 10 to 12 percent free silica, a known carcinogen. Therefore, dust exposure, expecially during underground blasting and ore crushing, should be minimized.
* In the retorting process, polynuclear aromatic hydrocarbons, heterocyclics, and other potentially carcinogenic compounds may be encountered.

Coal Gasification

* Phenol and carbon monoxide are emitted from the fluidized-bed and cyclone separator. Phenol is a known carcinogen, and carbon monoxide is toxic. An activated carbon adsorber and a flare can be used to control phenol emission.

- At the shift-conversion, there is potential exposure to hydrogen cyanide (HCN), which is highly toxic. Sodium hypochloride (NaOCl) with a strong caustic solution can be used to control cyanide emission.

Coal Liquefaction
- In the slurry mixing step, naphtha solvents are used.
- In the reactor, the slurry is dissolved, forming a highly aromatic liquid.
- The light-cut liquid products from the fractionator contain naphtha.

These substances are well known toxics and should be collected by using the control equipment indicated above.

4

Combustion of Fossil Fuels

The burning of fossil fuels, especially coal and fuel oils, presents problems of NO_x emission and the release of toxic hazardous substances. This chapter discusses NO_x formation, boiler NO_x reduction, and internal combustion (I.C.) engine emission control.

NO_x FORMATION

NO_x includes all oxides of nitrogen, primarily NO and NO_2. Field test data have shown that over 90 percent of the NO_x formed from combustion is NO, which oxidizes and forms NO_2 upon leaving the stack.

Two types of NO_x formation are associated with the burning of fossil fuels: thermal NO_x and fuel NO_x. Thermal NO_x formation is the thermal fixation of nitrogen in combustion air. Fuel NO_x is the NO_x converted from chemically bound nitrogen in fuel.

Almost all NO_x emissions from the burning of natural gas and light distillate oil are due to thermal fixation. When coal, residual oil, or crude oil is burned, the contribution of fuel NO_x to total NO_x emissions can be significant.

Thermal NO_x

The oxidation of nitrogen in combustion air was proposed by J. Zeldovich to be a chain reaction. Oxygen in combustion air is atomized:

$$O_2 + N_2 \rightarrow O + O + N_2$$

The oxygen atom reacts with nitrogen to produce nitrogen moNOxide, NO, and a nitrogen atom, N:

$$O + N_2 \rightarrow NO + N$$

The nitrogen atom reacts either with oxygen or with a hydroxyl radical, OH, at peak temperature, to form NO and either an oxygen atom or a hydrogen atom, respectively:

$$N + O_2 \rightleftarrows NO + O$$

$$N + OH \rightleftarrows NO + H$$

When the concentration of oxygen atoms reaches equilibrium, the atoms will combine and return to the molecular state:

$$O + O + N_2 \rightleftarrows O_2 + N_2$$

Thermal NO_x formation is directly proportional to the nitrogen (N_2) concentration, the residence time, the square root of the oxygen (O_2) concentration, and the exponential of temperature:

$$[NO] \sim [N_2] \cdot t \cdot [O_2]^{1/2} \cdot \exp(-1/T)$$

where
- $[\,]$ = mole fraction
- t = residence time
- T = temperature

Therefore, thermal NO_x can be reduced by the following tactics:

1. Reduction of the nitrogen concentration.
2. Reduction of the residence time or the time of exposure to peak temperature.
3. Reduction of the oxygen concentration.
4. Reduction of the peak temperature.

By lowering the volume of excess air, both N_2 and O_2 concentrations are reduced. In order to reduce the time of exposure to peak temperature, flue gas recirculation (FGR) is employed (see below, section on boiler NO_x reduction). This will reduce the O_2 concentration in the burners, causing flame temperature and peak temperature zone reductions, so that the fuel residence time at peak temperature is reduced. Several staged combustion methods are used to lower the local O_2

concentration, and the air preheating level is reduced to lower the peak flame temperature.

Mixing the fuel, air, and recirculating flue gas may increase or decrease NO_x formation. Increasing the swirl may increase the entrainment of the cooled combustion product so that the peak temperature is lowered. It may increase the fuel-to-air mix so that the intensity of local combustion is increased. Thus, various parameters of the system will affect NO_x formation.

Fuel NO_x

The nitrogen content of fossil fuels is reported to be as follows:

Fossil fuel	*Percent of N_2 content by weight*
Subbituminous coal	0.5–1.5% (dry, ash-free)
Bituminous coal	1.5–2.5% (dry, ash-free)
Crude oil	0.25% plus
Natural gas	5%

Most of the fuel nitrogen in natural gas is converted into NO_x in the combustion process, whereas 20 to 90 percent of the fuel nitrogen in oil or 5 to 60 percent of the fuel nitrogen in coal is converted to NO_x.

Of the fuel nitrogen atoms converted to NO_x, approximately 70 percent are vaporized, reacting with a free oxygen atom to form nitrogen moNOxide (NO) and a nitrogen atom:

$$N_2 + O \rightarrow NO + N$$

The nitrogen atom is oxidized to form NO:

$$N + O_2 \rightarrow NO + N$$

Vaporized fuel nitrogen is sensitive to the fuel-to-air ratio, is moderately sensitive to temperature, and reacts rapidly in NO_x formation.

The following most frequently employed methods of fuel NO_x reduction are discussed below: low excess air firing, optimum burner design, staged air combustion, staged fuel combustion, and secondary air preheating.

Low Excess Air Firing

In normal combustion, the amount of air supplied exceeds the calculated theoretical value to ensure the complete combustion of fuels. The excess air is expressed as a percent of the theoretical value; for natural gas: 5 to 10 percent; for fuel oil: 8 to 15 percent; for coal: 10 to 40 percent. On average, 20 percent excess air can be expected, depending upon the fuel conditions, such as the size of the fuel particles, the viscosity, the content of impurities, and the design of the fuel-burning equipment. At optimum conditions, the excess air can be reduced to a minimum so that NO_x formation also is minimized.

Optimum Burner Design

Burners are designed in different ways based on the types of fuels used (natural gas, fuel oil, and coal). For natural gas burning, several gas streams are injected into the discharge area of the burner, where combustion air is drafted around the gas streams. An effort is made to increase the mixing of gas and air, and dampers can be used to control the air supply.

For fuel oil burners, compressed air or steam can be used, to mix with the oil and to atomize the droplets. The oil also can be broken up mechanically into a fine uniform spray. Burners can be designed to maintain a local fuel-rich condition so that N_2 volatilization is minimized; NO_x formation also is kept to a minimum.

For coal-fired burners, the coal usually is pulverized, conveyed with air, and fed to the furnace. Combustion air is induced through the ports to the furnace. The relative locations of air and fuel entry, and the manner in which air and fuel are introduced into the furnace, are essential aspects of the burning operation. The pulverized coal may be fed vertically downward in a round nozzle or a long narrow slot, or horizontally through deflectors and vanes in the horizontal nozzle. Proper control of the coal burning rate, the air flow rate, and coal quality is essential to optimum design of the burner and the furnace. Turbulent mixing of the fuel and air will promote better burning conditions. Also the degree of pulverization will affect the coal burning efficiency, and dampers usually are used to control furnace conditions.

With high burning efficiency and low air consumption, NO_x emissions can be reduced. Further discussion of NO_x reduction is included below in the section on boiler NO_x reduction.

Staged Air Combustion

Air supplied to a burner is separated into two stages: primary and secondary air. In the first stage, insufficient air is supplied, causing incomplete combustion; therefore, the peak temperature is low. In the second stage, sufficient air is supplied to ensure that combustion is complete, as well as to cool the combustion gas.

Staged Fuel Combustion

Fuel is injected into two zones of a burner. In the primary zone, a portion of the fuel is injected into the combustion air and is burned lean (see below, section on internal combustion engines). Therefore, the flame temperature is low. The remaining fuel is injected downstream of the primary zone, completing the combustion at a low peak temperature.

Secondary Air Preheating
In staged air combustion, secondary air is preheated before entering the burner. This can have two effects:

1. More complete nitrogen volatilization (which maintains the nitrogen in a gaseous state) may occur, and less nitrogen may remain in the residue to be oxidized in the fuel-lean secondary stage, so that there is less NO_x formation.
2. The preheating of the air may increase thermal NO_x formation.

The use of this method should be carefully considered.

BOILER NO_X REDUCTION

NO_x-polluting equipment includes a wide range of devices, such as water heaters, ovens, furnaces, boilers, and so forth. To illustrate NO_x reduction from such devices, boilers are discussed in this section.

In order to control NO_x formation from a boiler, the function of the boiler and its components must be known, as well as the control technology. The function of a boiler is to generate steam at pressures above that of the atmosphere. Steam is generated by the absorption of heat produced in the combustion of fuel.

A boiler consists mainly of the following components:

- Drums, headers, and tubing that convert incoming feedwater into steam.
- A furnace in which combustion takes place.
- Burners that maintain combustion by mixing combustion air and fuel.
- Fans that supply combustion air and vent exhaust gases.
- An ash-handling system for coal burning.

Vertical tubes are connected at upper and lower headers, forming waterwalls. Downcomer tubes connect the bottom of a drum and the lower headers, and water is supplied from the drum through the downcomer tubes to the lower headers, from which water flows upward in the waterwalls. The waterwalls absorb heat from the furnace, and steam, along with a large quantity of water, is discharged from the top of the waterwall tubes into the upper headers and then passes through riser tubes to the drum. The water is separated from the steam in the drum and is returned to the waterwalls with the incoming feedwater. The steam can be superheated and may be used for power generation.

NO_x emission can be reduced by modifying boiler combustion. Topics discussed below include pulverized coal-fired boiler NO_x reduction, stoker coal-fired boiler NO_x reduction, oil-fired NO_x reduction, and gas-fired NO_x reduction.

Pulverized Coal-Fired Boiler NO_x Reduction

In this type of boiler, pulverized coal and air are blown into a chamber where they are mixed and burned in suspension. NO_x emission can be reduced substantially by modifying pulverized coal-fired boilers. Such modifications include using less excess air, putting burners out of service, overfiring the air injection, recirculating the flue gas, using low-NO_x burners, injecting ammonia, and reducing the firing rate. The effectiveness of such control measures and the environmental side effects are summarized below.

Lowering the Excess Air
The amount of combustion air is reduced from the conventional 20 percent excess air, and NO_x emission may be reduced from 0 to 25 percent. When the quantity of excess air is reduced to 5.2 percent, additional carbon monoxide (CO), hydrocarbons (HC), and smoke are generated in the exhaust gas.

Putting Burners out of Service
By cutting off the combustion air supply to one or more burners and maintaining the fuel-rich firing state (see below, section on I.C. engines), NO_x reduction of 27 to 39 percent can be expected. This modification can be made only on a boiler with four or more burners. A drawback is an increase in slag and corrosion.

Overfiring by Air Injection
Overfiring is accomplished by injecting (secondary) air through air ports located only above the fuel-rich firing burners, to reduce the peak temperature. NO_x reduction of 5 to 30 percent can be anticipated, but increases in slag and corrosion are a concern.

Flue Gas Recirculation (FGR)
One portion of the flue gas is recirculated to air-admission ports (a wind box) to make up 10 to 12 percent of the combustion air. NO_x reduction is from 0 to 20 percent. Additional costs are incurred, for installation of FGR ducts, a fan, and so on. FGR may cause combustion instability; modifications of the burners and the wind box are needed.

Low-NO_x Burners
Low-NO_x burners are designed by using controlled air–fuel mixing to meet the criteria for those burners. The criteria are expressed either on a weight scale as pounds of NO_x produced per one million Btu of heat delivered, or as NO_x

concentration in parts per million by volume at 3 percent oxygen content. Conversion from a measured NO_x concentration to a reference NO_x concentration at 3 percent O_2 is done by using the following formula:

$$(ppm)_{3\%} = \frac{17.9}{(20.9 - \text{dry \% of } O_2)} \times (ppm)_{measured}$$

Presently the criteria for a low-NO_x burner are defined as:

- 0.04 (lb NO_x/10^6 Btu) or 30 ppm for gas- and liquid-fired burners.
- 0.12 (lb NO_x/10^6 Btu) or 90 ppm for coal-fired burners.

These criteria can be changed, depending on advances in NO_x control technology. At present, 45 to 60 percent NO_x reduction has been achieved.

Ammonia Injection

Ammonia is injected into flue gas to decompose NO_x, forming N_2 and H_2O. NO_x reduction of 40 to 60 percent has been achieved for the equipment thus treated. This method can be applied better to newly designed equipment than for retrofitting existing boilers. Ammonia injection rates are limited to 1.5 NH_3/NO to avoid NH_3 emission.

Reduced Firing Rate

The fuel supply and the air flow to the boiler are reduced. NO_x emission has been found to be affected, in a range between a 45 percent reduction and a 4 percent increase in NO_x. The increase is due to an increase in oxygen resulting from operating at an off-design firing rate.

Stoker Coal-Fired Boiler NO_x Reduction

Stoker coal-firing is done to burn coarse solid fuels in a bed at the bottom of a furnace. Stokers are designed mainly to provide for continuous or intermittent fuel feed and the disposal of noncombustible materials. Based on the angle of fuel feed at the fuel bed, stokers are either underfeed or overfeed types.

In underfeed stokers fuel and air travel upward in the same direction, whereas in overfeed stokers the fuel enters the combustion zone from above traveling downward, in a direction opposite that of the air flow. Overfeed stokers throw solid fuel over from the distributor onto the bed so that a portion of the fuel burns in suspension, with the remainder burning on the bed. The bed can be a traveling, vibrating, or stationary grate.

Combustion modification techniques for NO_x reduction include lowering of the excess air, staged air combustion, reduction of the firing rate, reduction of the air preheat, and ammonia injection.

Low Excess Air

The air flow supply is reduced under the stoker bed. NO_x reduction may reach 5 to 25 percent, but a corrosion problem and high CO emission must be anticipated.

Staged Air Combustion

The air flow from under the grate (primary air) is reduced, and the air flow at the overfire air ports is increased. This will achieve a 5 to 25 percent NO_x reduction. A corrosion problem may occur, and CO emission will increase.

Reduced Firing Rate

There is a reduction of the coal and the air feed to the stoker. Changes in NO_x emissions may vary from a 49 percent reduction to a 25 percent NO_x increase. The boiler becomes less effective with a reduced firing rate.

Reduced Air Preheat

The combustion air temperature is reduced from approximately 470°K to 450°K. No_x emissions are reduced by 8 percent, and boiler efficiency also is reduced.

Ammonia Injection

Ammonia is injected in the convection section of the boilers, and a 40 to 60 percent NO_x reduction can be expected. This method should be implemented mainly on newly designed boilers. If it is used in retrofitting existing boilers, operational problems may occur.

Oil-Fired NO_x Reduction

Fuel oil must be converted to atomized particles before its combustion in burners, which consist of atomizers and registers. Fuel oil is atomized through the atomizers by using either pressurized fuel oil or pressurized steam. Registers are used to supply air to the burners.

Combustion modifications for oil-fired boilers for NO_x reduction include lowering the excess air, staging the combustion air, putting burners out of service, using flue gas recirculation (FGR), using FGR plus staged combustion air, reducing the firing rate, using low-NO_x burners, ammonia injection, and reducing the air preheat.

Low Excess Air

NO_x emissions can be reduced up to 28 percent for residual oil and 24 percent for distillate oil. Carbon moNO_xide, hydrocarbon, and smoke emissions may increase.

Staged Combustion Air
Burners employ fuel-rich firing at the initial stage and then have secondary combustion air ports to supply enough air to cool them down at the peak temperature. A 20 to 50 percent NO_x reduction for residual oil and a 17 to 44 percent NO_x reduction for distillate oil firing can be anticipated. This method works well on new boilers, but its use with a retrofit is impossible.

Putting Burners out of Service
The air supply to one or more burners is shut down, with the remainder allowed to fire fuel-rich. A 10 to 30 percent NO_x reduction for residual oil firing can be expected. No data are available for distillate firing. Use of the method for retrofit is accompanied by boiler derating.

Flue Gas Recirculation
A 15 to 30 percent NO_x reduction for residual oil firing can be achieved. A reduction of 58 to 73 percent has been reported for distillate oil firing. Fuel gas recirculation is best suited to new units; retrofitting existing units with FGR can be costly. Flame instability may be encountered at a 15 percent or greater FGR rate.

FGR plus Staged Combustion
This technology combines FGR and staged combustion. The NO_x reduction rate is 25 to 53 percent for residual oil firing and 73 to 77 percent for distillate oil firing. Retrofitting with this combined technology may not be possible.

Firing Rate Reduction
There is a reduction of air and fuel flow to the burners. NO_x emissions may be affected, ranging from a 33 percent decrease to a 25 percent increase for residual oil firing and from a 31 percent decrease to 17 percent increase for distillate oil firing. This method is not effective when excess oxygen is needed.

Low-NO_x Burners
The burners are designed with controlled air–fuel mixing and increased heat dissipation. For both residual and distillate oil firings, a 20 to 50 percent NO_x reduction can be expected.

Ammonia Injection
For both residual and distillate oil firing, a 40 to 70 percent NO_x reduction is anticipated. This technique is costly, and ammonium sulfate, $(NH_4)_2SO_4$, builds up; therefore, frequent cleaning of the burners is necessary.

Reduced Air Preheat

Combustion air bypasses the air preheater so that the temperature of the combustion air is reduced to 340°K, the ambient condition. NO_x reduction of 5 to 61 percent has been reported for residual firing, whereas for distillate oil firing no information is available.

Gas-Fired NO_x Reduction

Natural gas-fired boilers have achieved NO_x levels as low as 0.04 lb/10^6 Btu, levels that most low-NO_x burners for oil-fired boilers are incapable of reaching. However, the low-NO_x burners developed for oil-fired boilers can be used on gas-fired boilers to obtain additional NO_x reductions. Therefore, switching to natural gas is the most successful means of reducing NO_x emission.

INTERNAL COMBUSTION ENGINE EMISSION CONTROL

In addition to fossil fuel burning boilers, I.C. engines are major sources of NO_x emissions. this section discusses the types of I.C. engines, emissions from I.C. engines, and NO_x control of I.C. engines.

Types of I.C. Engines

Four-stroke cycle engines and two-stroke cycle engines commonly are used. The four-stroke types have a spark ignition system, and the two-stroke types have a compression ignition system.

The four-stroke cycle includes these strokes:

- Intake stroke: The engine sucks in the air and fuel mixture.
- Compression stroke: The mixture is compressed by the piston.
- Ignition and power stroke: A spark ignites, and the mixture burns, pushing the piston back to its initial position prior to the compression.
- Exhaust stroke: Exhaust gases are expelled from the cylinder.

The two-stroke cycle includes first and second strokes, behaving in the following manner:

- The first stroke includes simultaneous air intake and exhaust gas expulsion, followed by compression and fuel injection.

  *A blower blows air into the cylinder through windows at the center part of the cylinder. The exhaust gas is pushed out through the valve opening at the cylinder head.

*When the piston passes through the windows, the valve is closed, and the compression process begins.

*Fuel injection occurs, after which the first stroke is completed, and the second stroke takes place.

- The second stroke is the power stroke. The fuel is injected into highly compressed air so that ignition and combustion occur. The piston is pushed back to the initial position to complete the cycle.

Emissions from I.C. Engines

The type of fuel used for four-stroke cycle engines is mainly gasoline, whereas fuels for two-stroke cycle engines are diesel fuel, natural gas, sewage gas, and their mixtures. Emissions are NO_x, hydrocarbons, carbon moNO$_x$ide, particulate matter, and a small amount of SO_x, if a low-sulfur fuel is used.

From gasoline-powered engines, the following trace elements are detected: benzo(a)pyrene ($C_{20}H_{12}$), phenanthrene ($C_{14}H_{10}$), pyrene ($C_{16}H_{10}$), anthracene ($C_6H_4(CH)_2C_6H_4$), naphthalene ($C_{10}H_8$), and chrysene ($C_{18}H_{12}$). Benzo(a)pyrene is an active carcinogen, and phenanthrene, pyrene, and anthracene are suspected carcinogens.

From diesel-powered engines, in addition to benzo(a)pyrene, smoke and odor are emitted in the exhaust gases.

NO_x Control of I.C. Engines

Various methods have been tried to control NO_x emissions from I.C. engines. These techniques include derating, retarded ignition timing, air-to-fuel ratio adjustment, use of a turbocharge with aftercooler or intercooler, reduced manifold air temperature, exhaust gas recirculation, water induction, combustion chamber redesign (two-stage combustion), the use of catalytic converters, and a combination of the above methods. Their specific effects on emission control are summarized in the following paragraphs.

Derating
Engines can be operated by using a smaller supply of fuel (diesel engine) or by using an air–fuel mixture. The pressure and the temperature in the cylinder are reduced, thus lowering NO_x formation.

Retarded Ignition Timing
Spark discharge or fuel injection (diesel engine) is initiated when the piston is away from the top dead-center point. Then the combustion process is extended further into the power stroke and the exhaust period. In this way NO_x formation is decreased, but fuel consumption is high, efficiency drops, and backfiring

occurs in the extreme case. HC and CO emissions are insensitive to retarded ignition timing.

Air-to-Fuel Ratio Adjustment

In a stoichiometric relationship, the oxygen in the air–fuel mixture should completely oxidize the fuel. However, the oxygen content in the stoichiometric state has been determined to be 1 percent by volume of oxygen in the exhaust gas.

In the lean-burning state, the air–fuel ratio is higher than in the stoichiometric state. Therefore, the exhaust gas from a lean-burning engine contains more than 1 percent by volume of oxygen; NO_x emissions will be higher, but HC and CO emissions are decreased. In the rich-burning state, the air–fuel ratio is lower than in the stoichiometric state. The exhaust gas from a rich-burning engine contains less than 1 percent by volume of oxygen; NO_x emissions will be lower, but HC and CO emissions are increased.

If the air-to-fuel ratio is changed to the rich-burning state, NO_x formation will drop sharply, there being insufficient oxygen for combustion; and HC and CO emissions will increase accordingly.

Turbocharge with Intercooler

Here the exhaust gases from an engine are guided to a turbine that drives a compressor. The compressor compresses the intake air, which is cooled by an intercooler. Because of the lower intake air temperature and compression, a greater mass of air can be packed into the air manifold. Given the conditions of higher pressure and greater air mass, more fuel can be injected, to produce more power for a given size of engine.

The low temperature of the incoming air leads to a low peak temperature and thus to low NO_x emissions. It has been reported that 10 to 30 percent NO_x reduction has been achieved for diesel engines, but no data are available for gasoline engines. Because the turbocharge increases the air–fuel ratio to that of a lean-burning mixture, HC and CO emissions are decreased.

Reduced Manifold Air Temperature

The manifold air temperature can be reduced by using an intercooler upstream of the manifold. Because of the low air temperature, the air density is high, and a greater air mass can be packed. By injecting more fuel, the engine can produce more power. When there is a low air temperature, the peak temperature also is low; therefore, NO_x emissions are low. When the air temperature is too low, the combusion reaction is slow, so that HC and CO emissions increase.

Exhaust Gas Recirculation (EGR)

This control method involves recirculating the exhaust gas to replace one portion of the incoming air. External EGR is the recirculation of exhaust gas to the air manifold, whereas internal EGR is the restriction of one portion of the exhaust gas from exiting the cylinder. Cooling the exhaust gas and recirculating it to the intake will reduce the peak temperature, thus lowering the NO_x emissions. HC and CO emissions increase, because of a lack of available air for combustion.

Water Induction
Water is introduced into the engine either with the intake air or by water injection directly into the cylinder. The water is vaporized and the peak temperature reduced; thus, NO_x emissions are low. At low temperatures, hydrocarbons are burned slowly; therefore, HC emissions increase. CO emissions are unaffected by water induction.

Combustion Chamber Redesign
Combustion chambers are modified to have a cavity at the piston head. Fuel is injected into the cavity as a rich mixture and ignited. The mixture burns in the absence of excess air; thus NO_x formation is delayed. The burning mixture enters the main chamber, and is mixed with additional air, to complete the combustion and lower the peak temperature. Therefore, NO_x emissions are low, as are HC and CO emissions.

Catalytic Converter
Three-way conversion (TWC) catalysts are used to reduce NO_x, HC, and CO. Precious metal catalysts are used to oxidize HC and CO, and rhodium catalysts convert NO_x to N_2. By using a TWC catalyst, NO_x, HC, and CO can be reduced simultaneously.

5

Toxic Waste Management

This chapter discusses persistence and toxicity of pollutants, environmental impacts and governmental responses, toxic waste disposal technology, and developing toxic waste control technology. A summary of the five major toxic pollutants found in the coastal region of Southern California and their sources also is presented.

PERSISTENCE AND TOXICITY OF POLLUTANTS

When fossil fuels are consumed, some pollutants can easily be measured, but others cannot because the amounts of the chemical constituents are so tiny that they can be identified only by referring to evidence of their presence. These trace materials are chromium, arsenic, lead, cadmium, chlorine, fluorine, mercury, and so on, totaling approximately 25 to 30 elements. These elements are highly toxic and are distributed in fly ash, in bottom ash, or in vapor form. The worst locations are found in the vicinity of industrial areas where the heavy metals accumulate in soils or organisms near industrial sites.

Reportedly, some trace elements are transported by ocean winds to wilderness areas, forests, or mountains at some distance from the source. The concentrations of these trace elements in soils far exceed the normal background levels. Trace-element pollutants thus are widespread and accumulate everywhere in soil, sediment, or other organisms. Later they pass through the food chain and enter the human body. Even if the cause of this toxicity is removed, its effects persist.

Trace elements are both toxic and hazardous. Toxic materials are lethal, nondegradable, and biologically magnified, and they have detrimental cumulative effects. Hazardous materials pose potential dangers to human health and safety and to other living organisms in the environment. Toxic materials need not be hazardous if handled properly, but most hazardous materials are toxic because they pose a threat to human safety.

Trace elements can be divided into several categories: carcinogens, teratogens, mutagens, oncogens, and others. A carcinogen causes cancerous diseases; a teratogen causes abnormal growth of body organs; a mutagen can cause birth defects; an oncogen is a tumor-causing agent; other agents may cause nerve damage or sudden death. Therefore, the trace elements are classified as toxic hazardous materials. These materials can be identified by their toxicity, flammability, explosiveness, radioactivity, and acidity. A few examples are given in the following paragraphs.

Toxicity

This classification includes inorganic toxics and synthetic organic toxics. Inorganic toxics are metals, acids, and bases from chemical industries; synthetic organics are pesticides and polychlorinated biphenyls (PCBs), among others.

Flammability

Dusts or fine powders of organic substances (cellulose, flour), white phosphorus, films, volatile organic solvents (benzene), gasoline, circuit board manufacturing solvents (trichloroethylene), hydrogen, carbon monoxide, and other hydrocarbon gases fall into this category.

Explosiveness

Dynamite, ammonium nitrate prills (or pellets), TNT (trinitrotoluene), 20 percent ammonia in air, nitric acid fumes, and sulfuric acid fumes with water are explosive substances.

Radioactivity

Examples of radioactive materials are uranium and radium, which emit alpha-rays (helium-4), and radon gas. Uranium, alpha-rays, and radon gas are health hazards (see Chapter 10).

Acidity

Sulfur dioxide, hydrogen sulfide, sulfuric acid, and hydrogen fluoride are examples of highly acidic substances.

ENVIRONMENTAL IMPACTS AND GOVERNMENTAL RESPONSES

Hazardous materials affect the environment in many ways:

* They contaminate surface water and groundwater.
* They kill fish and livestock.
* They destroy wildlife habitats.
* They contaminate the soil.
* They damage crops.
* They cause fire, explosions, and air pollution.

In response to these negative effects, both federal and state governments have adopted bills designed to protect the public.

Federal Responses

RCRA (Resource Conservation and Recovery Act), 1976
RCRA is a supplement to the Clean Air Act (1963) and the Water Pollution Control Act (1972). It was adopted in 1976 with the following provisions:

* It provides the Environmental Protection Agency (EPA) with the authority to regulate waste disposal on land and solid waste management.
* It provides programs for research on the collection, separation, recovery, and recycling of resources.
* It establishes a cooperative effort among agencies to recover valuable materials and energy from solid waste.

RCRA is concerned mainly with technical considerations. Protection of the public from toxic substances and the revenue sources to support these activities are, if anything, merely implied in the act. Later, the Toxic Substances Control Act (TSCA) and the Comprehensive Environmental Response, Compensation, and Liability Act (CERCLA) were adopted to fulfill these objectives.

TSCA (Toxic Substances Control Act), 1976
TSCA was enacted in 1976. Its purpose is to protect human beings and the environment from exposure to chemical substances that may cause injury. Its main provisions are as follows:

* EPA must develop data to determine which chemicals are produced nation-wide, in what volumes, and with what risks.
* EPA must develop test procedures to determine whether chemicals presently in use are potentially harmful. If a chemical poses an unreasonable risk, its use can be limited or banned.

- EPA must be notified of new chemicals. Premarket evaluation of these chemicals will be performed to determine if they are carcinogens, oncogens, mutagens, teratogens, or other toxics (see preceding section).
- EPA publishes *TSCA Chemicals-in-Progress,* a bulletin used to report TSCA regulatory actions.

CERCLA (Comprehensive Environmental Responses, Compensation, and Liability Act), 1980

Enacted in 1980, CERCLA also is known as the Superfund Act. It mainly includes the following provisions:

- It provides the federal government with the authority to deal effectively with uncontrolled releases of hazardous substances to the environment.
- It assigns responsibility to federal and state governments to respond to and clean up hazardous substances and other toxic materials in the water, in the air, and on land.
- It creates a superfund by means of a tax imposed primarily on the chemical and petroleum industries.
- It forces private parties responsible for the release of toxic hazardous waste either to clean up or to reimburse the government for the costs of cleanup.
- It imposes requirements on members of the private sector to report to EPA as soon as they have knowledge of toxic releases.

With all the revenue and provisions included to protect the public, more specific steps were required to implement what Congress intended. These provisions are reflected in the RCRA—1984 Amendment.

RCRA—1984 Amendment

RCRA was amended in 1984 mainly to impose deadlines by which EPA would do the following:

- Identify hazardous wastes.
- Ban the use of landfills for untreated hazardous wastes.
- Clean up waste sites located by facilities, using the required permits.

From 1980 when CERCLA was enacted until 1985, a great deal of time was spent on clarification of CERCLA, but no cleanup activity was undertaken. After the Superfund Amendment and Reauthorization Act (SARA) was adopted in 1986, cleanup activities began and spread nationwide.

SARA (Superfund Amendment and Reauthorization Act), 1986

SARA was enacted in 1986. Its provisions include the following:

- It requires responsible parties to conduct response actions.

- It increases funding for underground tank cleanup.
- It imposes a mandatory schedule on EPA to expedite cleanup activities

Because of RCRA, SARA, and other legislative action, public concern about toxics in the air from chemical release has been increased, and further federal responses have developed.

"Hot Spots" Information and Assessment Act, 1987
Because of (a) confirmation that chemical manufacturing plants, facilities, and businesses using hazardous materials release substantial amount of toxic substances, and (b) the possibility that such releases may create localized concentrations or "hot spots" that present significant risks to individuals and population groups, the Air Toxics "Hot Spots" Information and Assessment Act was imposed in 1987.

The act mandated that:

- Information concerning amounts, exposures, and health effects of hazardous substances released from specific sources be collected and evaluated.
- The amounts and types of hazardous materials released from specific sources be ascertained and measured, and their health risks be assessed.

The "Hot Spots" Information Assessment Act soon was integrated as a part of California state law.

Responses from the State of California

California Environmental Quality Act (CEQA), 1970
CEQA mandated that all state agencies must prepare or cause to be prepared by contract an Environmental Impact Report (E.I.R.) before approving a project. An E.I.R. mainly shall include:

- An environmental impact statement about the project including:
  1. The present environmental status.
  2. Types and amounts of pollutants expected from the project.
  3. Future effects of the (newly produced) pollutants.
- Nonavoidable effects of the project.
- Mitigation measures and energy consumption reduction steps.
- Alternatives that might replace the original project.
- A statement about the relationships between short-term uses of the environment and long-term productivity.
- Information on irreversible environmental changes, including toxic waste management.
- A logical conclusion about the project's future impact.

Assembly Bill No. 2588, Connelly (AB 2588), 1987, Air Toxic Emission Assessments and Plans
The act requires the following:

- The state must enact the Air Toxic "Hot Spots" Information and Assessment Act.
- The state must compile a list of substances that present a chronic and acute threat to the public.
- The operators of air toxic–emitting facilities must prepare and submit a proposed comprehensive emission inventory plan for review and approval by the local air pollution control agency by a specified date.
- As an alternative, the local air pollution control agency must prepare an industry-wide emission inventory and risk assessment for any class of facilities that meet certain criteria.
- Within 180 days after approval of the plan, the facilities must implement the plan and report on it to the local air pollution control agency.
- The air pollution control agency must review the report and notify the State Department of Health Services, the Department of Industrial Relations, or city or county health departments of its findings and determinations.
- The facilities must update the emissions inventory biennially.
- After review of the emission inventory data, the air pollution control agency must prioritize and categorize the facilities for purposes of health risk assessment into high, intermediate, and low categories.
- Within 180 days of categorization, the operator of every high priority category facility must prepare and submit to the air pollution control agency a health risk assessment using scientific methods.
- In case of a significant health risk, the operator must notify all affected persons about the health risk.
- The air pollution control agency must prepare and publish an annual report summarizing health risks.
- The state must use the report to identify and control toxic air contaminants.

All these responses should be kept in mind during the following discussion of major existing and developing technologies related to toxics disposal.

TOXIC WASTE DISPOSAL TECHNOLOGY

All trace elements, or toxic hazardous pollutants, should be controlled at the source. They must be collected and removed before they are released and spread throughout the environment. The most desirable means of control is to recycle the hazardous by-products and return them to use. The majority of toxic wastes, however, are converted into harmless or less harmful substances. Some of the wastes are put in permanent storage, which is the least desirable control method.

Conversion of Toxic Hazardous Pollutants

The methods of converting toxic hazardous pollutants include incineration, thermal destruction, biological treatment, chemical destruction, land farming, and ocean assimilation.

Incineration

In this process toxic air contaminants make direct contact with an open flame; the combustible wastes are burned and changed into gases and ash. The combustion temperature can be as high as 2000°F although a catalytic bed can be used to lower the combustor temperature.

Incineration is very effective in breaking down complex organic compounds, such as pesticides, solvents, and polychlorinated biphenyls, into gases, or in detoxifying them. It is the safest method of hazardous waste disposal, the waste volume is reduced, small land areas are required, and the heat produced can be used for heating or for power generation. However, the cost of incinerating toxic waste is very high, including waste transportation and fuel consumption expenses. Also, wastes are not all combustible, and incineration of explosive substances is banned. Incomplete combustion of certain hazardous wastes renders them more dangerous than the original wastes if they are not controlled properly.

Incineration at sea is cheaper than incineration on land, and can minimize the danger to the public. Some concerns of this method are chemical spills, accidents due to human error or poor weather conditions, and the residue associated with incomplete destruction of toxic waste, which threatens marine life. A cost–benefit analysis must be performed before any method is selected.

Thermal Destruction

Thermal destruction employs non-flame-contact thermal energy to destroy toxic hazardous wastes. It includes catalytic oxidization, fluid-bed combustion, the pyrolytic process, and other methods.

Catalytic Oxidization

Toxic gases are guided into a device with a catalyst bed installed in their path so that the gases are oxidized as they pass through the bed before being vented into the atmosphere. The catalysts used are platinum and oxides of metals such as TiO_2, CuO, or AgO.

An example is a catalytic oxidizer used to control an ethylene oxide (EtO) sterilizer, EtO being a suspected carcinogen. After passing through a hopcalite catalyst, a mixture of copper and nickel oxides, EtO is converted to water vapor and carbon dioxide.

Fluid-Bed Combustion

A fluid-bed combustor consists mainly of a vertical reactor. A certain amount of powder (coal, limestone, or sand) is placed inside the reactor, with a start-up burner submerged beneath the powder. Liquid waste is mixed with combustion air, which is blown into the reactor in the vicinity of the burner. The powders are blown upward and form a homogeneous mixture, with the temperature and composition evenly distributed. The mixture becomes a fluid bed where both liquid and solid wastes are burned as soon as they are fed into the reactor. The bottom ash is removed by the release of a screw at the lower end of the reactor.

To control the temperatures of the fluid bed and of the exhaust gases from the reactor, water is sprayed on the fluid bed. The exhaust gases are vented through a cyclone (Appendix A), a waste heat boiler, and a scrubber, and then into the atmosphere.

Pyrolytic Process

The pyrolytic process is based on the chemical decomposition of organic substances by heating them in anaerobic conditions. Pyrolysis of organic materials results in three classes of products:

1. A solid residue containing char, carbon, and ash.
2. A condensable liquid, including water and certain organics.
3. A gas mixture of methanol, ethanol, and isobutanol.

This method is the most promising means of treating refuse.

In 1972, the following chemical analysis of raw refuse was reported by J. J. Mikovich:

Composition	Percent by weight
Moisture	20.00
Carbon	29.83
Hydrogen	3.99
Oxygen	25.69
Nitrogen	0.37
Sulfur	0.12
Ash and metal	20.00

Mikovich also reported the products that follow a pyrolytic operation:

Products	Percent by weight
Inorganic removed (during process)	9.00
Moisture removed (during process)	20.00
Charcoal	11.25
Liquids	40.80
Gas	7.95
Ash and metal	11.00

An application of the pyrolytic process for waste destruction is the Landgard system, which extracts combustible compounds and oxidizes them. Refuse is fed into a rotary kiln (Chapter 6), where the flame temperature is raised to approximately 1500°F at the inlet. Hydrocarbon gases and moisture are extracted and guided to a combustion chamber for oxidization of the hydrocarbons. Then an afterburner is used to complete the combustion. Exhaust gases are ducted to gas scrubbers for acidic or caustic treatment before being released into the atmosphere. The spent water from the scrubbers is pumped into a water clarifier to separate the water from salts. The water is used in a water quenching unit to cool the charcoal residue leaving the kiln. The quenched residue is conveyed through a magnetic unit for removal of ferrous materials, and the remaining wet residue is transported to a landfill.

Other Thermal Destruction Processes
Other processes of thermal destruction use molten salt, plasma, and microwave reactors:

- A salt in the molten state, also known as a fused salt, has a temperature of 500 to 1000°C. When toxic hazardous wastes are guided through a molten salt bed, the toxic substances are destroyed.
- Plasma is a stream of positively charged ions that exist only at very high temperatures, ranging between 40 million (40×10^6) and seven billion (7×10^9) degrees Celsius. Toxic substances that are guided through the plasma are destroyed, for all electrons are stripped from the atomic nuclei.
- A microwave is an electromagnetic wave with a wavelength between one and 100 cm. Microwaves penetrate into chemical compounds, producing enough heat to detoxify specific materials.

Of these three processes, the molten salt reactor process is the one currently used in industry. The other two, plasma and microwave processes, have been developed and laboratory-tested, but are not yet commercially available.

Biological Treatment
Biological treatment consists of using microorganisms to decompose certain pollutants and remove toxic elements. Microorganisms can be grouped into three categories: heterotrophs, photoautotrophs, and chemoautotrophs.

Heterotrophs
This group of microorganics, also known as heterotrophic bacteria, utilizes organic carbon compounds as a carbon source (or as building blocks) and an energy source. It converts protein or cyanic acid into carbon dioxide, water, and nitrogen, as shown in the following reactions:

$$C_6H_{12}O_6 + 6O_2 \rightarrow 6CO_2 + 6H_2O$$

$$4HOCN + 3O_2 \rightarrow 4CO_2 + 2H_2O + 2N_2$$

Photoautotrophs
This group of bacteria uses carbon dioxide as a carbon source and sunlight as an energy source. These photosynthetic bacteria mainly control heavy metals, such as cadmium, lead, and hexavalent chrome.

Chemoautotrophs
This group of bacteria also is known as nitrifying bacteria. It uses carbon dioxide as the carbon source, and the energy comes from the oxidation of inorganic compounds by nitrifying bacteria. The bacteria include *Nitrosomonas* and *Nitrobactor*. *Nitrosomonas* converts ammonia into the nitro radical (NO_2^-), and *Nitrobactor* converts the nitro radical (NO_2^-) into nitrate (NO_3^-):

$$2NH_3 + 3O_2 \xrightarrow{\textit{Nitrosomonas}} 2NO_2^- + 2H^+ + 2H_2O$$

$$2NO_2^- + O_2 \xrightarrow{\textit{Nitrobactor}} 2NO_3^-$$

Nitrate compounds can be absorbed by the root systems of green plants. Biological treatment is a slow process; it cannot be used to treat a stream dosage of concentrated toxic materials.

Chemical Destruction
Using chemical compounds to react with toxic materials to convert their structures into simple compounds is known as chemical destruction. An example is the conversion of toxic cyanide (NaCN) to less toxic cyanate (NaCNO), or toxic hexavalent chrome (Cr^{+6}) to less toxic trivalent chrome (Cr^{+3}):

$$\underset{\text{cyanide}}{NaCN} + Cl_2 \rightarrow NaCl + CNCl$$

$$CNCl + 2NaOH \rightarrow \underset{\text{cyanate}}{NaCNO} + H_2O + NaCl$$

$$\underset{\text{hexavalent chrome}}{4CrO_3} + 3Na_2S_2O_5 + 3H_2SO_4 \rightarrow 3Na_2SO_4 + \underset{\text{trivalent chrome}}{2Cr_2(SO_4)_3} + 3H_2O$$

Land Farming
This method combines biological and chemical means of treating toxic elements. Approximately one foot of topsoil is mixed with the waste; then chemical and biological reactions decompose part of the waste, and the remaining waste is dilute enough for its migration to be harmless to the environment.

Several requirements should be met in land farming:

- The site environment should be conducive to bacterial activity.
- The site area must be well plowed, and waste material should not overload the area.
- The selected site must have runoff-water control.
- The site must be well above the groundwater level to avoid water contamination.

In some areas, land farming is not permitted because of air pollution problems.

Ocean Assimilation

Certain wastes can be distributed in the ocean without harming human health. Many wastes have been dumped into the ocean, including: dredging spoils from rivers or lakes; industrial wastes, such as heavy metals, cyanide, mercuric compounds, and chlorinated hydrocarbons; sewage sludge; construction and demolition debris; refuse; and explosive materials. These wastes affect the ocean water quality, which must be kept at a constant, acceptable level to protect marine organisms. The average standard of ocean quality is as follows:

- Salinity: approximately 3.5 percent salts average.
- Temperature: approximately 40°F average.
- pH concentration: in the 7.8 to 8.2 range.
- Dissolved oxygen: sufficient amount to support ocean life.
- Turbidity: at a level such that photosynthetic reactions in aquatic organisms are not prohibited.

Substances that are not allowed to be discharged into the ocean are toxic substances, petroleum products, settleable solids, floating objects, and nuisance organics.

Permanent Storage

Those toxic wastes that are not treated are disposed of permanently in storage sites. Storage methods include disposal in landfills, underground injection, surface impoundment, salt formation, and arid region burial.

Landfills

There are problems associated with toxic waste disposal in landfills. Primarily, landfill sites may be phased out within the foreseeable future. Also, questions of long-term legal liability, landfill gas migration, toxic air pollution, and toxic liquid waste evaporation are main concerns in landfill disposal.

A double liner structure has been proposed for toxic waste landfill use. Underneath the waste material, four bottom layers are provided in the following sequence: the leachate collection layer, the first impervious liner, the leachate detection layer, and the second impervious liner. Groundwater monitors also are included.

Underground Injection
Toxic liquid wastes frequently are injected into idle gas or oil wells. They are far below groundwater level and do not cause immediate harm to the groundwater. However, their future effects on the ecosystem are not completely understood.

Surface Impoundment
Toxic wastes are gathered and mixed with fly ash and cement and encapsulated with plastics. The capsules are being stored aboveground awaiting future disposal.

Salt Formation
Toxic wastes are being dumped in abandoned salt mines, in the hope that the wastes will be converted to salt in years to come.

Arid Region Burial
In this method, toxic wastes would be buried under a water-free desert in an unsaturated zone. Although this method has been suggested, it is not yet being practiced.

DEVELOPING TOXIC WASTE CONTROL TECHNOLOGY

Some toxic waste control technologies have been developed successfully on a laboratory scale, but have not yet become commercially available. These processes include waste reduction, volume reduction, chemical detoxification, fixation, and stabilization.

Waste Reduction

Waste reduction can be achieved only by changing processes, changing raw materials, or eliminating a process or production line. Examples of waste reduction are given in the following paragraphs.

Changing Process Conditions
In one chemical manufacturing process, during the steam cracking (Chapter 9) of naphtha to yield ethylene at high temperature and high pressure, phenol, a suspected carcinogen, is produced and is found in the condensate water. If the

process conditions are changed to those of low temperature and low pressure, no phenol is obtained.

Modifying Process Procedures
In the process of phenol chlorination, dioxin is produced as a toxic by-product. To avoid dioxin buildup, it is necessary to purge the system regularly. Addition of the purge operation to the process can minimize the buildup of toxic by-products.

Changing Feedstocks
Instead of cracking naphtha to produce ethylene, ethane can be fed to the process for cracking. With this method, the by-products are relatively pure, and little or no phenol is entrained in them.

Volume Reduction

This process is the removal of nonhazardous materials from toxic waste so that the volume of the toxic waste can be reduced. Methods include separation, concentration, and precipitation.

Separation
Toxic wastes are manually separated from nontoxic materials. The handling costs may increase, but analysis and disposal costs will decrease.

Concentration
Moisture is distilled or evaporated from liquid waste. Because concentration methods are energy-intensive, the operations can be very costly.

Precipitation
For nonsoluble or minimally soluble salts, such as lead sulfate ($PbSO_4$), a filter is used to separate liquid from solid matter.

Chemical Detoxification

By using chemical reactions, molecules of hazardous materials can be converted to those of nonhazardous materials. For example, a chlorinated hydrocarbon can be dechlorinated by a chemical reaction. Products of the reaction will be a nonchlorinated hydrocarbon and chlorinated salt.

Specifically, chloroform ($CHCl_3$), a suspected carcinogen, can react with potassium hydroxide (KOH), to produce potassium formate (HCOOK) and potassium chloride (KCl). The reaction is shown below:

$$CHCl_3 + 4KOH \rightarrow HCOOK + 3KCl + 2H_2O$$
chloroform   potassium formate

The product, potassium formate, is a salt of formic acid and can be collected by boiling dry.

Fixation

The addition of chemical substances to toxic waste to prevent rapid evaporation is known as toxic fixation. Fixing agents include cement, fly ash, lime, and blast furnace slag. After fixation, the hazardous compounds, such as organic matter salts or other soluble salts, become less volatile and nonleachable.

Stabilization

Substances are added to toxic wastes to stabilize them. For example, some mixtures have been converted into roadbrick. In this case, the materials are encased and cannot migrate in the environment.

SOURCES OF TOXIC POLLUTANTS IN THE COASTAL REGION OF SOUTHERN CALIFORNIA

The top five toxic pollutants that are suspected carcinogens in the Southern California coastal region are:

- Formaldehyde, CH_2O
- Benzene, C_6H_6
- Methylene chloride, CH_2Cl_2
- Trichloroethylene, $CHCl:CCl_2$
- Ethylene oxide, CH_2CH_2O

Their sources and control measures are summarized below.

Formaldehyde

The major sources of formaldehyde are: photochemical reactions of nitrogen oxides and hydrocarbons emitted from plastics and resin manufacturing, combustion processes in the petroleum refining industry, diesel trucks and automobiles, and chemical plants.

The process used to control formaldehyde is not widely accepted; the general approach is to use incineration at a temperature of approximately 1600°F. Activated carbon adsorption (Appendix B) is not recommended for the control of formaldehyde emissions. Its drawback is heat adsorption, which can produce tar that plugs the carbon bed, causing a foul odor and a fire hazard.

Benzene

Benzene is emitted primarily from refineries, asphalt plants, solvent usage, gasoline stations, and pharmaceutical manufacturing processes. Benzene emissions are controlled by using a flare or an incinerator at approximately 2000°F. An activated carbon bed is not recommended.

Methylene Chloride

The sources of methylene chloride are degreasers, paint strippers, blowing agents in foam manufacturing, and pharmaceutical and chemical plants. The control technology for methylene chloride includes a refrigerated freeboard chiller, an activated carbon adsorber, or incineration at approximately 2000°F.

Trichloroethylene

The sources of trichloroethylene are degreasers, the use of adhesives for Plexiglas, rubber cement manufacturing, polyvinyl chloride production, and refrigerants. The control technology for trichloroethylene includes a refrigerated freeboard chiller, an activated carbon adsorber, and incineration at approximately 2000°F.

Ethylene Oxide

The sources of ethylene oxide emissions are sterilization of medical equipment, meat products, coffee, spices, and hospital uses. The control equipment for ethylene oxide includes catalytic oxidizers, fluid-bed oxidizers, catalytic afterburners, chemical scrubbers, and sulfuric acid–impregnated activated carbon adsorbers.

The toxic pollutants discussed above are either directly or indirectly related to energy resources, and particularly to environmental concerns. Further exploration of the control of toxic pollutants and their management is ongoing in the industrialized nations.

6

Recycling of Solid Waste

INTRODUCTION

Wastes are by-products of modern society, which have been treated and disposed of for generations. Instead of being thrown away, some solid wastes are recycled successfully and cost-effectively. The feasibility of recycling is illustrated by the general cycle of matter discussed below. The separation of wastes produces recyclable solid wastes that can be categorized in different groups, and can be treated appropriately. The remaining residues can be either disposed of in a landfill or burned for heat energy recovery. Thus, recycling saves energy and, in some cases, provides energy resources.

General Cycle of Matter

The concept of the general cycle of matter is closely associated with the law of conservation of matter, which says that matter is neither created nor destroyed but simply changed from one state to another. This implies that matter somehow can be recycled.

In contemporary society, there is an abundance of matter for industrial, commercial, and residential use. The resulting wastes are transported to recycling plants where reclaimable materials are recovered and are furnished to the society together with other natural resources for consumption. The nonreclaimable materials are waste residues that can be burned, producing heat energy that can be used by consumers. An alternative is to bury the waste residues, a method that produces landfill gas, a natural energy resource.

Separation of Waste

Wastes that can be recycled include municipal and industrial wastes. The wastes are transported to a recycling plant, where they are dumped and loaded on a primary conveyor and then transferred to a separation conveyor, where bundled paper, paper products, and cardboard are manually transferred to a paper salvage conveyor. Hazardous materials also are removed manually. The remaining wastes are gravity-fed into a pulverizer unit, which reduces the waste size by a hammermill shredder. Underneath the shredder is a sizing screen, which holds the larger-sized waste for crushing by the hammermill. The smaller-sized waste falls down through the screen grits along a vertical duct to a belt conveyor.

At the center section of the vertical duct, a horizontal cross duct is connected, and a fan is installed facing the horizontal direction inside the horizontal duct, so that fine particles falling down from the screen can be blown to the horizontal duct. These light fractions are collected for recovery of combustible refuse (see following section).

Before the belt conveyor transfers the sized fractions into a hopper, a magnet-separating conveyor intercepts ferrous materials and transfers iron or steel wastes to a separate hopper. The first hopper collects nonferrous materials, and the second hopper contains ferrous materials. The nonferrous materials are guided to vibrating screens for further separation of glass, aluminum, copper, and rubber tires.

REPROCESSING AND EMISSION CONTROL

Paper Recycling

Topics discussed below include the following: the sources of recycling materials, the air pollution problem, the water pollution problem preceding the pulping process, and the water pollution problem during the pulping process.

Sources of Recycling Materials
Materials suitable for paper recycling are newspapers, magazines, spent papers, cardboard, packing materials, corrugated materials, and other paper products. After they are collected, they are either shredded and wired into dense bales or compacted in large containers to be transported to a processing plant.

The advantages of paper recycling are that it saves trees, saves energy, and reduces air pollution from paper mills. However, paper recycling also has disadvantages: it causes water pollution from the deinking of printed papers and the bleaching of wastewater from pulp mills; it does not carry a tax write-off; and it decreases the amount of material burned for heat energy recovery.

The Air Pollution Problem

There are some air pollution problems associated with paper recycling. Newspapers are printed with solvent-based inks, which evaporate during recycling and mix with moisture in the air. The mixture possesses an unpleasant odor, which creates a nuisance problem. Therefore, moist papers must pass through a dryer before they can be processed in paper mills.

Several control methods have been used to solve the odor problem (Appendix B) of the dryer. They include a venturi scrubber, an electrostatic precipitator (Appendix A), a filtration system, and incineration. However, the operating costs for a venturi scrubber are very high; the oil coating on the electrodes of an electrostatic precipitator may decrease the collection performance; activated carbon filters may be expensive; and the fuel costs of incineration are extremely high, especially for a high moisture content air mixture.

The Water Pollution Problem before the Pulping Process

Before entering the pulping process, the wastepaper must be deinked and bleached. Normally the basic coloring materials, dyes, are positively charged organic molecules, and the fibers are prepared to contain negative sites that attract and interact with the positively charged molecules. The basic dye structures are triphenylmethane, $(C_6H_5)_3CH$; phenazine, $C_6H_4N_2C_6H_4$; and triazine, $NC(OH)NC(OH)NC(NH_2)$. The decoloring can be done by using an aqueous solution containing enough alkali to adjust the pH from 8 to 10. The solution is heated to approximately 190°F and held there for approximately one hour. Then the bath is cooled and discharged, and fresh water is added to rinse the fibers.

Depending upon the chemical constituents and the degree of saturation of the pigments, the methods of removing the color from the wastewater differ. If a small amount of color is to be removed, it can be most efficient to use biological treatment with sludge separation; the pigments will be adsorbed on the sludge and removed from the water. If the concentration of the pigments is high, biological treatment may not remove much of the color; then control equipment should be used.

The control equipment includes a carbon adsorber, a chemical coagulator, and an HEPA (high efficiency particulate air) filter, depending upon the type of pigment to be removed. A carbon adsorber is not suited for a disperse type of pigment, and chemical coagulation is not suited for an acid pigment. The HEPA-filter produces good results, but at a high cost.

The Water Pollution Problem during the Pulping Process

After deinking and bleaching, the wastepaper will be reprocessed into pulp at paper mills. Normally the wastepaper will be placed inside a digester and decomposed by heat and chemicals. In the past, calcium bisulfite was used as a cooking chemical in paper mills, but the spent cooking acids were disposed of in rivers so that in time waste dotted the riverbanks. Other pollutants include suspended solid pulp, fiber, and gaseous emissions of sulfur dioxide (SO_2) and hydrogen sulfide (H_2S).

The gaseous emissions are flare-burned, and the liquid wastes are treated physically and biologically. The physical treatment, or primary treatment, consists of skimming the floating part and removing the sediment. The biological treatment, or secondary treatment, consists of adding activated sludge to cause precipitation, which can be removed.

The effluent discharge from the cooking chemicals was prohibited by law in the 1970s, necessitating a production change. A new pulping method was developed to take care of the water-pollution problem; the basic chemicals presently in use are magnesium hydroxide ($Mg(OH)_2$) and sulfur (S).

Sulfur is burned to become SO_2, which is supplied to a digester with $Mg(OH)_2$. The digester contains SO_2, $Mg(OH)_2$, wood chips, and wastepaper. Its temperature is raised to approximately 325°F, and its pressure is maintained at 90 psig for 6 to 7 hours. Then the pulp is blown to a perforated wooden pit where the spent cooking acid is drained through the perforated openings to a storage tank, and the pulp is transferred to a paper-making process.

The process includes the following steps: The pulp first is washed and then is vacuum-filtered. The cleaned fiber passes over screens sized for different grades of pulp, and debris and dirt are removed. The fibers are thickened to increase their constituents and bleached to brighten their color. The fibers are arranged on a plate, water is removed by pressing, and the sheet is heat-dried. Additional heat is supplied to the sheet when it passes through a paper machine, and, after chemical coating, the final product is complete.

The spent cooking acid is a weak liquor containing 9 percent solid matter. The weak liquor is transferred to an evaporator to condense it to a heavy liquor of 60 percent solids content. The heavy liquor then is burned in a boiler to produce steam. The flue gas contains SO_2 gas and magnesium oxide powder (MgO), which is separated from the gas stream by a cyclone. Magnesium hydroxide can be produced by injection of steam into magnesium oxide:

$$H_2O + MgO \rightarrow Mg(OH)_2$$

Both sulfur dioxide gas and magnesium hydroxide can be recycled as basic chemicals, and the cycle of the pulping process can be started anew.

Combustible Refuse Recovery

One of the most promising solid waste disposal processes is the pyrolysis method, which employs high-temperature (approximately 1500°F) breakdown of organic matter in anaerobic conditions (where oxygen is absent) to generate liquid and gaseous fuels from organic waste. This method has been used to recover fuels from the light-fraction refuse.

In fuel recovery, the refuse is sealed in a furnace that is electrically heated to approximately 1500°F. Hot gas is driven off and is guided away from the furnace for further treatment; the remaining refuse is char residue, which also can be used as fuel. The hot gas is piped through an air-cooled trap and is condensed to produce heavy oil and tar. The remaining vapor is guided to a water-cooled liquor, which usually contains an aromatic solution. The noncondensable oil mists are further collected by an electrostatic precipitator, and the exhaust gases are acid- and caustic-treated in chemical scrubbers prior to being burned in a flare.

Typical yields of fuel recovery per ton of municipal waste have been reported as follows:

- Char residue: approximately 200 pounds at 10,000 Btu/lb.
- Heavy oil and tar: approximately 5 gallons.
- Light oil: approximately 2 gallons at 150,000 Btu/lb.
- Organic gas: approximately 15,000 cubic feet at 500 Btu/ft^3.
- Aqueous liquor: approximately 100 gallons.
- Ammonium sulfate: approximately 20 pounds.

Ferrous Material Recycling

Ferrous materials recovered from waste separation are transferred to an electric furnace, melted into ingots, and then sent to a rolling mill to be formed into bars or angle irons. Three types of electric furnaces are available: direct-arc, indirect-arc, and induction furnaces.

Electrodes of direct-arc furnaces are submerged in the ferrous materials. When alternating current is charged to the electrodes, arcs are formed between the electrodes. Radiation heat is released from the arcs, and resistance heat is produced within the bath that melts the ferrous materials. The electrodes of an indirect-arc furnace, on the other hand, are exposed to the top of the ferrous materials. Radiation heat from the arc and convection heat from the pool of molten materials will melt the freshly loaded materials.

An induction furnace consists of a line of refractory materials inside the furnace wall and an electrical coil surrounding the outside of the wall. All these elements are placed in a welded steel casing. After ferrous materials are charged in the furnace, alternating current is switched on to the coil, which induces an eddy current in the charged materials. The resistance heat of the alternating current melts the ferrous materials.

The advantage of using an electrical furnace is that the furnace temperature is controlled and can be as high as 6000°F, the melting point of ferrous materials.

The percentage of pure material also is high. The disadvantage is the high cost of electricity, especially for large-volume operations.

The effluent from ferrous material recycling must be cooled prior to being guided to the control equipment. The criteria for control equipment selection depend mainly upon the regulatory requirements.

For the least stringent requirements, a scrubber with an alkaline solution is used to remove particulate matter and sulfur dioxide. This device can handle high temperatures and sticky gases, and takes up only a small plant space. Additionally, it has a fire protection system. The drawback of using a scrubber is that the water has a high solids content, which must be treated properly. In addition, a large amount of water is required, and corrosiveness and duct damage are important considerations.

For moderately stringent regulations, a baghouse (Appendix A) can be used. Baghouses can handle various dust particle sizes, dust loadings, and flow rates. Because bag fabrics are sensitive to abrasion, moisture effects, and chemicals, dust characteristics must be known prior to final fabric selection.

For the most stringent regulatory criteria, an electrostatic precipitator can be used. The advantage of using an electrostatic precipitator is that its collection efficiency is high, the pressure drop across the equipment is low, and it is capable of handling moisture and oily substances. However, the equipment is sensitive to the flow rate, the flow direction, the temperature of the flue gas, and the moisture content. Also, a cyclone is needed to collect large particles upstream of the electrostatic precipitator, and the problem of disposing of the dry dust that is collected must be solved.

Nonferrous Material Recycling

The nonferrous material residue from solid waste separation is transferred to vibrating screens, where glass, aluminum, copper, and rubber tires are separated for further treatment. The recycling of these materials is discussed in the following paragraphs.

Glass Recycling
Glass, porcelain, and ceramics are collected, washed, separated by color, crushed, and transported to glass manufacturing plants. These types of waste can be used to make new glass, mixed with asphalt to be used for paving roads, or added to bricks and cinder blocks for construction purposes.

Glass can be classified into five categories:

1. Soda-lime glass: 70 percent silica (SiO_2), 15 percent soda (Na_2O), and 10 percent limestone (CaO).
2. Lead glass: 70 percent silica and 15 percent lead (Pb).

3. Borosilicate glass: 80 percent silica and 15 percent boric oxide (B_2O_3). It has a very low expansion coefficient and transmits ultraviolet radiation.
4. Ninety-six percent silica glass: 96 percent silica and 4 percent boric oxide.
5. Silica glass: 99.8 percent silica with minimal impurities.

Glass products are flat glass (25%), containers (50%), tableware (10%), and miscellaneous glasses such as equipment (15%).

Waste glass can be recycled in a batch plant. The broken glass (cullet) is fed through a magnetic separator and then to a crusher. The cullet is dropped by gravity to a receiving hopper and is transferred through a screw conveyor and a bucket elevator to a rotating head. The rotating head distributes the cullet and other major compounds, such as silica, soda ash, and limestone, in separate compartments of a bin. Other minor ingredients, such as lead and boric oxide, are stored in different compartments of another bin.

All compartments are guided to a weighing hopper so that each compound can be measured to meet specifications. After being weighed, the compounds are gravity-fed to a mixer; and the emissions from the mixer and the hopper are controlled, for example, by using a baghouse, prior to being released into the atmosphere.

The mixed materials are transported through a belt conveyor to a bucket elevator, which carries them to a storage bin. The materials are fed through a feeder to a glass-melting furnace. The glass furnace consists mainly of two chambers: a melter and a refiner. Burners are installed at one side of the chamber, and raise the glass temperature to approximately 2700°F. The molten glass flows from the melter chamber to the refiner chamber through a submerged opening between them. The glass temperature at the refiner chamber is reduced to approximately 2300°F; the molten glass can be poured through the forehearth to a glass-forming machine.

The emissions from a glass-melting furnace are due mainly to combustion and glass melting in the furnace. These emissions are particulate matter and gaseous compounds. The particulate matter includes potassium oxide (K_2O), sulfuric anhydride (SO_3), calcium oxide (CaO), boric anhydride (B_2O_3), silica (SiO_2), and many other compounds. The gaseous compounds include nitrogen, oxygen, water vapor, carbon dioxide, and a small quantity of nitrogen oxides. The major air pollution problem thus is due to the particulate matter, which can be controlled by proper handling of the raw materials and careful attention to fuel combustion practices.

Fine particles of raw materials are easiest to process in the melting stage; but because of dust reentrainment, dust carryover and particulate matter emission are high. Therefore, there is some cutoff point for particle size in an optimum melting operation.

Sometimes water is sprayed on the fine materials to reduce emissions. Up to

2 percent by volume of water spraying is reported to be optimal for emission reduction. Excessive water spraying will wet the particles, causing them to adhere to conveyors, elevators, storage bins, and other devices.

Particulate matter from melting furnaces frequently is controlled by baghouses, which have filters made of silicone-treated fiberglass. Exhaust gases are cooled to 400°F, which is above the dew point (280°F) of SO_2, to avoid condensation. The efficiency of a baghouse is approximately 99 percent.

Centrifugal wet scrubbers are used to control a high flue gas flow rate, but the efficiency of this type of control is quite low (52%). From a fuel-usage standpoint, large furnaces are more economical than small ones; the corresponding emissions are less than those from small furnaces. A rapid change in the raw material feed rate will waste fuel and increase emissions. The air-to-fuel ratio should be automatically controlled by using an oxygen trim.

Aluminum Recycling

Sources of aluminum waste include residues of molten aluminum molds, casting metal returned from foundries, and aluminum scrap. Because the scrap contains impurities or a variety of alloy metals, unwanted metals must be removed from the molten metal by fluxing. The fluxing process uses chlorine gas or another such agent, which is blown up from the bottom of a molten bath so that oxides and dirt may be removed from the molten metals by skimming them out from the melt surface.

The aluminum reclamation process includes chipping, cleaning, loading, melting, fluxing, and pouring. Aluminum scrap is reduced in size by a chipper. Scrap pieces contaminated with oil, grease, or paints are burned clean in a chip dryer before being charged to the melter furnace.

For a small operation, say up to 1000 pounds per batch, a crucible furnace is used (see below). The operation will take 4 to 5 hours per batch. For a large operation, say between 2.5 tons and 50 tons per batch, a reverberatory furnace is used. This operation may last up to 72 hours per batch. In either case, it is common practice to melt heavier scrap first and to add lighter scrap below the melt surface to avoid oxidation.

A crucible furnace is a cylindrical shell of welded steel lined with refractory materials, such as clay–graphite mixture or silicon carbide, for fire resistance. A vertical control rod is attached to the furnace so that molten aluminum can be poured out from the top. Gas burners are installed tangential to the furnace near the bottom section, and exhaust gas is discharged from the opening at the center of the lid where the chipped aluminum is charged. The furnace temperature is near 2500°F.

A reverberatory furnace consists of a refractory lined chamber, a charging door, and wells. Fuel burners are installed in the chamber wall, and heat is provided by a luminous flame through radiation and reflection off the roof and wells onto the aluminum scrap. Exhaust gas is vented by a hood atop the furnace, and a chlorine gun is installed at the bottom of the well. Chlorine gas, aluminum chloride, or aluminum fluoride is fed under pressure through the gun, forming chlorine gas bubbles across the molten layer of aluminum. Dirt, oxides, dissolved gases, and magnesium can be removed by fluxing. The chlorine gases are collected by a submerged hood covered by the molten metals. To pour off the molten aluminum, the chlorine gun is used with the pressure removed.

The chip dryers and furnaces are sources of emissions. The emissions from chip dryers are smoke and odors, which can be controlled by using an afterburner. The peak temperature is 1400°F to 1600°F, with an exhaust gas retention time of 0.5 to 1.0 second. The emissions from furnaces are particulate matter and gaseous emissions, such as chlorine and fluorine. The control equipment can be caustic scrubbers or alkaline powder-coated fiberglass filters.

Caustic scrubbers may be packed tower or venturi types. The packed tower scrubber is very effective for halogen gas control, but it may become plugged when dust loading is high; the venturi scrubber will remove particulate matter and halogen gases simultaneously. Fiberglass filters coated with a powdered alkaline adsorbent can be used to control particulate matter and halogen gases at the same time. The coated material neutralizes the acid gases, and the filter collects fine particles. The collection efficiency can be high.

Copper Recycling

Copper scraps are collected from different sources and transported to a reclamation plant, where copper is separated from the attached items and is made molten in copper refining or reverberatory furnaces. Sources of copper scraps are wire, including a large quantity of power cables; household items; telephone wire; transformers/alternators; and other electrical equipment.

Copper parts are removed from each attached item by hand-cutting, sawing, or melting of soldered joints. Wire and cable are stripped by machines when cables are large enough to be handled mechanically. The insulation of small wire can be burned off in a multiple-chamber incinerator or a kiln.

An incinerator (which is a device in which combustible wastes are burned and changed into ash and gases through a regulated process) can be used for small operations of wire reclamation. A batch of scrap wire is loaded into an ignition chamber where primary burners supply the required heat. Exhaust gas is blown through a second chamber equipped with a secondary burner and a fan for complete combustion before being vented into the atmosphere. The purposes of using multiple chambers are to maintain the required temperature, to extend the residence time, and to obtain turbulent mixing of the gas flow. The reclaimed wire is ready to be fed into a furnace after the insulation is removed.

For a large amount of wire reclamation, a continuous rotary kiln can be used. A kiln is an oven, a furnace, or a heated enclosure, which usually has a long cylindrical housing that slopes from the feed entrance to the tail end of the cylinder. Inside the housing is a cylindrical rotatable furnace, where primary burners fire at the feed side, and combustion air is supplied from the tail end of the kiln. During the operation, batch loads continuously move from the feed entrance into the furnace; and because of its rotation and slope, the batch loads are skidded downward to the tail end of the kiln. The insulation is burned completely at the tail, and the batches are ready to be picked up by carts for further processing. The exhaust gases are vented from a stack located at the batch-feed entrance, and a secondary burner completes the burning at the stack before the exhaust gases are treated in a control device.

A rotary kiln can maintain a sufficient temperature, a long residence time, and good turbulence for continuous wire reclamation. However, air-pollution problems occur at the furnaces, which require control equipment similar to that used for aluminum recycling, for the burning of the insulation.

For a halogenated plastic insulation, in which polyvinyl chloride may be used, the emissions are hydrogen chloride, hydrogen fluoride, hydrocarbons, and particulate matters. Caustic scrubbers are used to control hydrochloride and hydrofluoride emissions. Hydrocarbons, which may include aromatic (containing benzene rings; see Chapter 9), aliphatic (containing open chains), olefinic (containing one double ring of C_nH_{2n}), and particulate matter, can be controlled by using an afterburner.

For non-halogenated-plastic insulation materials, such as cotton, silk, paper, or rubber, the emissions are carbon dioxide and water vapors. An afterburner operating at the following conditions is needed to control the hydrocarbons and particulate matter: a flame temperature from 1400°F to 1600°F and a residence time from 0.3 to 0.6 second at the primary chamber; and 1400°F and 0.7 second at the secondary chamber.

Rubber Tire Resource Recovery

After the treads are worn off, some tires are recapped, whereas others are used for other purposes. Ultimately, all of them are dumped in scrap piles. Tire scrap cannot be disposed of in landfills because of its resistance to biodegradation. Instead, two methods are used to recover the resources in rubber tires: a pyrolytic technique for combustible fuel and incineration for thermal energy recovery.

Scrap tires are shredded, and then they undergo the pyrolytic process, yielding gaseous, liquid, and solid compounds. The gaseous compounds contain more than 50 percent hydrogen with 900 Btu/ft^3 heating value. Liquid compounds include approximately 15 percent olefins by volume, 50 percent aromatics, and 35 percent paraffins and cycloparaffins combined.

Olefins are unsaturated open-chain hydrocarbons that contain at least one double bond and thus are photochemically reactive. Aromatics are unsaturated cyclic hydrocarbons containing at least one benzene ring, a six-carbon ring with three alternate double bonds. These compounds are rather highly reactive and are suspected carcinogens. Paraffins are straight carbon chain hydrocarbons that vary with increasing molecular weight from methane (a gas) to waxy solids. Cycloparaffins have three or more carbon atoms united in a ring, with each carbon joined to two hydrocarbon atoms (CH_2), such as cyclopropane (C_3H_6) or cyclobutane (C_4H_8). These compounds are narcotic, and there is little difference in effect between high and low concentrations, as they can cause narcosis and even death.

The remaining solid residue from the pyrolysis of tires is a mixture of carbon, sulfur, and ash, with a heating value of approximately 13,000 Btu/lb. The equipment used to control the emissions from the pyrolytic technique consists of an electrostatic precipitator, a flare, and caustic and acidic scrubbers.

The use of tire incineration for thermal energy recovery began in about 1974 in West Germany. The world's largest tire incinerator, however, is the Westley facility, located in Stanislaus County, California, which has been in full operation since 1988.

Scrap tires in the Westley facility are dumped from a 500-foot belt conveyor into two boilers. Steam generated by the boilers is fed into a steam turbine, which is coupled with a generator that produces 3000 megawatts/hour of electricity. Spent steam is condensed by a condenser, and waste heat is disposed of by a cooling tower. The burning rate for the tires is 3 tons/hour at each boiler, and the emissions are NO_x, particulate matter, and acidic gases. Ammonia is sprayed into the exhaust gas to remove nitrogen oxides, and then fabric filters are used for particulate matter removal. Following these processes, a caustic scrubber is used to remove acidic gases by contact with limestone ($CaCO_3$); the acidic gases react with limestone to produce gypsum, which is collected in a sump. The treated gases are vented through a stack into the atmosphere.

NONRECYCLABLE WASTE

After different wastes have been recycled, the remaining residue can be either buried in a landfill or converted to energy. The processes of landfill waste disposal and the handling of gases from landfills are summarized below. Then the arguments for and against the use of landfills, a risk assessment, and comments and a recommendation about the incineration of wastes are presented.

Landfill Gas Recovery

Processes in Landfills
After wastes are buried in a landfill, three kinds of processes generally occur: physical, chemical, and biological.

Physical processes include compression of wastes, settling the wastes at certain locations, breaking their ties with other items, moving them from one place to another, their absorption of the surrounding liquid, and their adherence to other elements.

Chemical processes mainly include reactions of chemical compounds. A common process is the reaction between acidic and metallic compounds. After the chemical reaction, the waste residue proceeds to the next step, which is a biological process.

Biological processes start from the fragmentation of organic compounds in waste residues. Through aerobic decomposition, organic compounds are converted to carbon dioxide and water. Then, after most of the surrounding air has been consumed, fragments of waste are dissolved in water, and anaerobic waste degradation begins. The organic material dissolved in water is converted to an alcohol and ammonia by bacteria. Then the alcohol is further converted to methane to complete the formation of landfill gas.

Gases from Landfills

After several steps in the conversion processes, landfill gases are formed. Although the heating value of these gases is low, their quality can be upgraded to a level that renders them useful as an energy source. The composition, yield, and upgrading of landfill gases are discussed in the following paragraphs.

Composition
Landfill gases consist mainly of methane, carbon dioxide, nitrogen, oxygen, and a trace amount of sulfides. The amounts of these compounds present are as follows:

- Methane (CH_4): 40 to 50 percent.
- Carbon dioxide (CO_2): 30 to 40 percent.
- Nitrogen (N_2): 10 to 20 percent.
- Oxygen (O_2): 1 percent.

Gas Yield
The yield of landfill gas is linked to the amount of organic materials in the waste: the greater the amount of organic material is, the greater the amount of methane. The maximum landfill gas yield is approximately 10 cubic feet for each pound of volatile organic solids. The production rate is gradually decreased to 10 percent of the maximum rate within about 5 years.

Upgrading
The heating value of landfill gas is approximately 200 to 700 Btu/ft^3, depending upon the organic content of the waste. This is low in comparison with the value for natural gas, which is 1050 Btu/ft^3. The quality of landfill gas can be upgraded

by removing moisture from it. One possibility is to expose landfill gas to an ethylene glycol–alcohol system for dehydration.

Waste to Energy

Pros and Cons
The conversion of waste (garbage) to energy is a controversial subject. The following is a summary of arguments both in favor of and against the technology employed to convert waste to energy by incineration.

Arguments Favoring Conversion of Waste to Energy
- All waste should be treated in one way or another, and solid waste problems need to be taken care of.
- The waste materials used to produce energy cost nothing.
- Landfills are overflowing, and communities are running out of landfill space. Besides, landfills create and transfer toxics to the groundwater.
- Waste is converted to ash by incineration; so the volume and the weight of the waste are reduced.
- In addition to the advanced treatment methods used, such as pyrolysis, composting (organic wastes are shredded, decomposed, and used as conditioner to increase soil fertility and porosity), or recycling of waste, two generic methods have proved to be feasible, as discussed below.

1. Modular Technology. A standard, compact unit is used, with prepared fuel firing, for light hospital use and plastic waste burning. The combustion temperature is from 1600°F to 1800°F; the waste capacity is limited to 1500 lb/day. When waste is burned, the exhaust gas is treated by using a thermo-de-NO_x method (see Appendix B), followed by a cyclone and then by a baghouse before ventilation to the atmosphere.

2. Mass Burn Technology. The Martin-Stoker technology is a proven refuse incineration process, which has been used to treat over 65,000 tons/day of trash worldwide. The Stapelfeld Refuse Incineration Plant, in Germany, has a capacity of 19 tons/hour of refuse (or 456 tons/day), and approximately a 150 million Btu/hour firing rate. The plant consists of a bunker, an overhead crane, a hopper, a moving arm, grates, a furnace, a water-tube boiler, an electrostatic precipitator, a flue gas washer, a caustic scrubber, and a stack.

Garbage is transported and dumped into a bunker. Then an overhead crane picks up the garbage and dumps it into a hopper. A moving arm pushes the garbage into moving grates, which shift trash into the combustion chamber of the furnace where the fuel and air are supplied for an initial burning. Flue gases pass through a water-tube boiler to generate steam. After the heat energy is transferred to water, the flue gases are treated by an electrostatic precipitator, a flue gas washer, and a caustic scrubber before being released to the stack.

Arguments against Conversion of Waste to Energy
- The burning of waste is not cost-free because it requires fuel.
- Garbage burning fouls the air.
- Controls are inadequate to take care of toxic pollutants.
- Motor vehicle traffic is already congested, and before burning the waste would be transported through crowded areas.
- The technology that has been developed is inadequate for disposing of the toxics created in the conversion.
- Failure of the existing technology would raise some important questions: What should be done? Who is going to pay for it? (A large amount of money would be involved.) Who will guarantee that required policies will be written with the interests and needs of small and middle-sized businesses in mind?
- Combustion products contain dioxin and furan, which are very toxic. How would they be disposed of?
- No standards have been established for dioxin and furan, which are new additions to the toxics list and should be handled carefully. We must not approve technology that results in the additional production of these toxics.
- The accumulative effects of toxics should be considered. They should not be thought of as having separate or single effects.
- In a nonattainment area, any additional pollution will make the air quality worse.
- The waste could be recycled; substitute materials should be developed to reduce the use of toxic substances.
- Technological, scientific, and safety knowledge of toxics should be acquired first, before the construction of conversion plants begins.
- Risk assessment indicates that the maximum individual risk is 12 to 36 in one million in such conversions. This is extremely high compared with the one-in-one-million ideal (see below).

Cancer Risk Assessment
A cancer risk assessment is made to assess the probability of cancer risk based on the findings of laboratory analysis or other scientific methods. This assessment includes the maximum individual cancer risk and the public health risk. A brief description of the risk assessment and a simplified risk screen method are presented below.

Maximum Individual Cancer Risk ($R_{\text{max. ind.}}$)
The maximum individual cancer risk indicates the highest probability of a person's becoming a victim of cancer caused by his or her individual exposure to a carcinogenic pollutant from a source. It is determined by a formula:

$$R_{\text{max. ind.}} = q \cdot c_{\text{unit}} \cdot U$$

where:

q = emission rate of a carcinogenic air contaminant, in g/sec
c_{unit} = unit concentration, in $(\mu g/m^3)/(g/sec)$
U = unit risk factor, in $1/(\mu g/m^3)$

The emission rate, q, can be determined by either source testing or analytical calculation. The unit concentration, c_{unit}, is the long-term average of the ground concentration of the contaminant with a 1 g/sec carcinogenic emission flow rate. (For a conservative but quick estimate, certain simplifying assumptions were made to precalculate a set of data in which the unit concentrations are listed as a function of the following parameters: stack height, distance to receptor, and 8 hours or 24 hours per day of operation.) The unit risk factor, U, is a dose–response factor showing the cancerous risk for a person exposed to 1 $\mu g/m^3$ of the toxic substance over a 70-year lifetime.

All the parameters must be determined and integrated in a computer model to estimate the incremental cancer risk for individuals affected by the subject source. Numerous computer models have been developed, including the Gaussian Air Quality, Industrial Source Complex, Simple-Terrain, and Complex-Terrain models. The specific purposes of these models are described in the following paragraphs.

The Gaussian model assumes normal distribution of pollutants at any cross section of a plume downwind of a source. The Gaussian equation for the ground-level concentration, C, is:

$$C = [q/(\pi)\, s_y \cdot s_z \cdot u]\, \exp\,[(-1/2)\,(Y/s_y)^2]\, \exp\,[(-1/2)\,(H/s_z)^2]$$

where:

C = pollutant concentration (g/m^3)
q = emission rate (g/sec)
s_y, s_z = diffusion parameters in the y- and z-coordinates, respectively (m)
u = wind speed (m/sec)
x, y, z = downwind, cross-wind, and vertical coordinates, respectively (m)
H = the elevation (z-coordinate) of the emission point (m)

The Gaussian model is designed to be applied to relatively stable pollutants for a short-term averaging period (say for a daily average).

The Industrial Source Complex (I.S.C.) model is a steady-state Gaussian plume model designed to assess pollutant concentrations from a wide variety of sources associated with an industrial source complex. It deals with aerodynamic downwash, particle deposition, and the separation of point sources, and it can be used for area sources and volume sources.

The Simple-Terrain models are used to simulate an area where terrain features are all lower than the top of the stack. A single-source model has been recommended to assess pollutant concentrations for short and long terms in rural areas. It is a steady-state Gaussian dispersion model designed for point sources at a single location in either a rural or an urban setting, and it has been used to evaluate most of the criteria pollutants.

The Complex-Terrain model is used for a terrain exceeding the height of the stack being modeled, and is intended mainly for particulate matter and SO_2 modeling. It handles plume impaction, flow separation around an obstacle, and the high concentration of toxics found in the shelter of the obstacle.

Public Health Risk

The public health risk is the overall health impact on an entire exposed population. It is usually expressed as excess cancer cases, an indication of the increase in the occurrence of cancer cases in a population subject to a lifetime (70 years) of individual cancer risk that is greater than one in one million. Numerically it is equal to the product of the population potentially exposed to an individual cancer risk of greater than one in one million for 70 years times the maximum individual risk. The sum of the total affected excess cancer cases should be kept below 0.5.

Multiple Pathway Risk Analysis

The method used to calculate a multiple pathway risk is referred to in the *Air Toxics Assessment Manual,* prepared and published by the California Air Pollution Control Officers Association (CAPCOA). In addition to the above-calculated maximum individual cancer risk (due to inhalation), other pathway risks are involved, such as vegetation, runoff water, and soil contact. These cancer risks can be categorized as the ingestion cancer risk, which is the product of the soil toxics concentration, the soil ingestion rate, a gastrointestinal absorption factor, and the ingestion potency slope per unit body weight.

Another approach is to estimate the total risk associated with a given inhalation risk. A multiple pathway adjustment factor, MP, is listed in the Technical Support Document for the "Report on Chlorinated Dioxins and Dibenzofurans" published by the former California Department of Health Services (DOHS). The total cancer risk is equal to the MP factor times the maximum individual cancer risk (due to inhalation).

Comments and Recommendation

The following comments pertain to the controversy over whether incineration is a viable method of disposing of waste (garbage):

- Little consideration has been given to policy determination.
- Incinerators handle only one-fourth of residential wastes.
- Landfills soon will be phased out. Landfill users need time to deal with this problem.
- Available sources for landfills have not been studied carefully.

- Commercial and industrial wastes are handled by private contractors, who do not reduce their volume.
- Alternatives to incineration and their consequences are not known.
- Plant sites are chosen in midtown or poor areas, where transportation arteries already are crowded, and the plants would make the traffic worse. Concerns in site selection include: property values, health effects, legal liabilities, the type of plant that should be built, and where and when it should be built.
- Fly ash contains polychlorinated dibenzo-*p*-dioxin (PCDD) and polychlorinated dibenzofuran (PCDF) (see following discussion).

PCDD and PCDF are highly toxic. It is known that they are initiators and promoters of cancer; upon accidental human exposure, they will cause liver damage, liver cancer, bladder cancer, central nervous system damage, and skin disorders. PCDD and PCDF are formed in the gas plume because some combustion products are chlorinated by free-radical chlorine present in the gases. The burning of chlorinated compounds polychlorinated biphenyl (PCB) and 1-1-1 trichloroethane will produce PCDD and PCDF.

Proper ways to control PCDD and PCDF include the removal of plastic products from the source and good combustion practices, including a sufficient combustion zone temperature (1800°F), a proper residence time (1 sec for PCDD, 4 sec for PCDF), turbulent mixing, and a proper air/fuel ratio. The control efficiency is approximately 99 percent (weight reduction).

7

Waste Heat Management

Most industrial processes are accompanied by waste heat; and because of environmental and economic concerns, the management of waste heat is being increasingly scrutinized. The major sources and dispersal methods for waste heat are discussed below, and then the environmental impacts of each cooling technique are considered. Cogeneration is included to illustrate the conversion of waste heat into useful energy.

SOURCES OF WASTE HEAT

According to the second law of energy, in any conversion of energy there is always a portion of energy that is less useful than the other energy of the process. This portion of less useful energy is the loss or waste of heat that exists in almost every process. The major sources of waste heat are power plants, paper mills, petroleum refineries, foundries, and garbage-burning municipal incineration facilities.

The efficiency of a power plant is very low. Only about one-third of its heat energy can be converted into electrical energy; so two-thirds of the heat input is waste energy. It has been estimated that in a fossil-fuel power plant, about 85 percent of the waste heat is carried off by cooling water, and 15 percent is lost in stack gases, in boiler heat transfer, and in the piping system. In a nuclear power plant, about 95 percent of the waste heat is carried off by the cooling water. Other statistical data indicate that paper mills produce 12 times more waste heat than the plant could possibly use, and petroleum refineries produce 28 times their own heat-energy needs.

Depending upon the process, exhaust gas temperatures differ, but they usually are very high. For a copper reverberatory furnace, the exhaust gas temperature ranges from 1650°F to 2000°F. Therefore, foundries and municipal incinerators also produce a large amount of waste heat.

This waste heat constitutes both an ecological and an economic problem. The following section indicates how waste heat is dispersed into the environment.

WASTE HEAT DISPERSAL METHODS

Methods for waste heat dispersal can be classified as once-through cooling, cooling ponds, cooling towers, and heat recovery through equipment and various operations. The characteristics of the individual methods are described below.

ONCE-THROUGH COOLING

A once-through cooling system usually is installed near a river where water flows year-round.

While fuel energy is supplied to a boiler, steam is generated and ducted to a steam turbine, which drives an electric generator that converts the energy into electricity. After waste heat is released to a condenser, the spent steam is condensed as water and returned to the boiler. The waste heat is removed by passing cooling water through the condenser, whose intake and discharge ends are connected to a river or an ocean to dissipate the waste heat.

Pertinent considerations are the water flow rate of the condenser, which should be adequate to carry off the condenser heat; the river flow rate, which should be sufficient to remove heat; and the distance between the intake and the discharge of the condenser, which should be properly chosen so that no discharged water is returned to the condenser. It is not always feasible to place a plant near an ocean or a river; an alternative is to build a small lake (a cooling pond) next to the plant.

COOLING PONDS

Building a cooling pond for waste heat removal has the advantage that the pond also can be used for water sports or for warm water fish culture. Because heat is removed through evaporation and radiation, a large area is required. Sometimes a pump is installed to spray water from the pond into the air to increase the water surface area for evaporation. This reduces the pond area, but energy consumption increases.

When a natural body of water is not available, and a cooling pond is not desirable, it may be necessary to build a system to do what a pond does but in less space. Then a cooling tower is used.

COOLING TOWERS

Cooling towers can be categorized as wet cooling and dry cooling towers. In wet cooling towers, waste heat is removed through air–liquid direct contact; in dry cooling towers, waste heat is transferred from the hot liquid through a piping wall to cold air. Wet cooling towers can be subdivided into these types: counterflow induced draft cooling tower, crossflow induced draft cooling tower, counterflow natural draft hyperbolic tower, and crossflow natural draft hyperbolic tower. Fiberglass-reinforced plastic also is of interest in cooling tower applications; so it is included in the following discussion.

Wet Cooling Towers

Counterflow Induced Draft Cooling Tower
A counterflow induced draft cooling tower is shaped like a rectangular wood box. The inside of the box contains fill packings; its top is a fan, and its bottom is a cold water sump.

Hot water is pumped from the condenser exit to the top of the tower plenum, and is sprayed downward over the fill packings. Air is drawn through a louver located at the tower base and moves upward to contact water droplets. After stripping the heat from the water, hot air is vented into the atmosphere. The (heat-stripped) cold water is accumulated at the sump and is pumped into the condenser for reuse.

Two types of fill packings are available: a splash type and a film type. In the splash type, the packing wood plates are installed horizontally, causing water to break up to create the required surface for heat transfer. In the film type, the thin sheet packing elements are spaced close together and suspended vertically. Water is distributed over the elements and runs over the sides to create a water film. The film surface area is the primary factor in sizing a cooling tower.

The surface area, A, can be calculated by a heat balance formula:

$$Q = m \cdot c_p \cdot (\text{range temp.}) = U \cdot A \cdot (LMTD)$$

where:
Q = heat recovered from water
m = water flow rate of condenser
c_p = specific heat of water
U = overall heat transfer coefficient

and *LMTD* is the log mean temperature difference, which is defined as:

$$LMTD = \frac{GTD - TTD}{\ln \dfrac{GTD}{TTD}}$$

The reference temperature is t_{ss}, the saturated steam temperature, which corresponds to the spent steam temperature at the entrance of the condenser. With t_{hw} and t_{cw} denoted as hot water and cold water temperatures, respectively, the grand temperature difference *(GTD)*, the terminal temperature difference *(TTD)*, and the range of temperature can be defined:

$$GTD = t_{ss} - t_{cw}$$

$$TTD = t_{ss} - t_{hw}$$

Range of temperature $= t_{hw} - t_{cw}$

The required heat transfer surface area, A, in the cooling tower can be calculated from the above formulae.

Crossflow Induced Draft Cooling Tower

A crossflow cooling tower has a rectangular box shape similar to that of a counterflow cooling tower. However, the air inlet louver is extended from the base up to the top of the tower, and water is pumped to the water basin at the top of the tower and flows vertically downward through the fill.

The fill packings usually are of a splash type; air is drawn horizontally into the tower and comes into contact with the splashing water droplets. Then the hot air is passed through turning vanes at the center of the tower and moves upward, to be vented into the atmosphere.

This tower has the advantages that the higher the water dispersion rate, the higher the heat transfer will be, and the longer the retention time, the better the heat transfer will be. The pressure drop across the cooling tower is minimal. Also, the total size of a crossflow induced draft cooling tower is less than that of the counterflow type because it can be operated at a high velocity. The system is preferred because of its small size, minimum pressure drop, and cheaper cost, compared with the counterflow induced draft cooling tower. Also, it is the more efficient design, because of its longer retention time.

Counterflow Natural Draft Hyperbolic Tower

The principle of the natural draft design is that the air temperature at the bottom of the tower is higher than the temperature at the top of the tower. This creates a difference in air density that induces air flow through the system. A hyperbolic curve is believed to provide the optimum curvature for the design of natural draft towers. Hot water is pumped to a plenum installed at the midsection inside the tower and is sprayed over fill packings. Warm air is drafted through the base into the tower and flows upward in contact with the droplets. Waste heat is transferred from the hot water to the air flow and is released through the top of the tower into the atmosphere. The cold water is accumulated in a sump and is pumped to a condenser for reuse.

Crossflow Natural Draft Hyperbolic Tower
Here the hyperbolic shell is constructed of reinforced concrete on the top of a hot water basin, and hot water is pumped from the condenser to the basin. The water flows downward, through several steps, to a cold water basin. Warm air is drafted across the downfall water stages, where waste heat is transferred to the air, which then moves upward through the tower to the atmosphere.

This type of hyperbolic tower can handle larger heat loads than the counterflow type and is used at large power plants. The wet cooling towers have certain problems, however, and some special characteristics are required of tower materials to overcome the problems. The structures must do the following:

- Resist corrosion.
- Minimize damage due to long-term exposure to water.
- Withstand different weather conditions.
- Prevent chemical attack.
- Reduce fungus growth.
- Eliminate either shrinkage or swelling of redwood-fill packing materials.
- Minimize the shutdown of tower cells for repair of the fan or other portions of the tower.

Fiberglass offers many of the needed characteristics. Also, because of its availability in different colors, it makes the towers aesthetically more acceptable.

Dry Cooling Towers
There is no direct contact between air and hot water in a dry cooling tower. While air is induced through the cooling tower, hot water is guided from a condenser to a piping system inside the tower. After the waste heat is transferred through the piping wall to the air, cool water is returned to the condenser.

Although no water loss is involved in using a dry cooling tower, its cost is increased significantly over that of a wet tower because of its poor cooling efficiency.

Heat Recovery through Equipment

Different types of equipment have been used to recover waste heat from different sources, including pipes and ducts, air preheaters, recuperators, regenerators, economizers, waste heat boilers, condensers, and heat exchangers. The configurations of these devices differ, but the principle of their design is the same, based on heat transfer phenomena. The operation of these devices is summarized below.

Pipes and Ducts

A hot gas exhaust duct and a cold air inlet duct are installed next to each other, and a bundle of pipes is passed transversely through the exhaust duct and the inlet duct. A heat transfer fluid inside the pipes absorbs heat from the hot gas and evaporates to form a vapor. The latent heat of vaporization is carried in the vapor to the cold end, where the vapor condenses and gives off its heat to the cold inlet air. The condensed fluid is pushed through the outside pipes of the bundle and back to the hot end, where the cycle begins again.

Air Preheaters

Air preheaters contain alternate channels: heating gas channels are installed vertically, and cold air channels horizontally. Heating gas and cold air are in close contact with each other, separated by a thin wall of conductive metal. Thus thermal energy is transferred from the heating gas to the cold air so it can be used as combustion air. These devices are commonly used in boilers in various industries.

Recuperators

Recuperators, which are shaped much like cylindrical towers, are used to regain a portion of the heat from exhaust gases released from furnaces. Exhaust gas is ducted from a furnace into the base of the recuperator and flows upward to the stack of the recuperator while fresh air is pumped to a coiled piping system inside the recuperator tower. After heat is transferred, the air is warmed and is supplied to burners as combustion air.

Regenerators

Regenerators are used to preheat natural gas before feeding it to the combustion chambers of gas turbines. A gas-line piping system is arranged horizontally in several layers inside the regulator, with a natural gas inlet end at the top layer and an outlet end to the gas turbine at the lowest layer. The exhaust gas from the turbine is ducted to the bottom of the regenerator; and after heat is transferred from it to the natural gas, the exhaust gas is released for further treatment.

Economizers

Economizers sometimes are used in boiler applications where cold water is heated before being fed to the boiler. The economizer may have the shape of a simple housing installed near the boiler exhaust. Inside the housing is a finned tube system that recovers waste heat from the exhaust gas and transfers it to cold water passing through the tube system to the water head of the boiler. The exhaust gas temperature can drop from 500°F to 300°F across the economizer.

Waste Heat Boilers

These boilers generate hot water or steam by absorbing heat from exhaust gas without burning additional fuel. Feedwater is circulated with the exhaust gas before entering the water drum of the boiler. This approach is economically sound and has been applied to various industrial processes.

Condensers

Hot exhaust vapors are guided into condensers through vapor tubes. After the latent heat is released to cold water around the vapor tubes, the vapor temperature will drop, from T_2 to T_1, and the cooled condensate is accumulated for further treatment. The hot water can be used for processes in chemical plants or petroleum refineries. The heat recovery by water is expressed as follows:

$$Q = m \cdot c_p \cdot (T_2 - T_1)$$

where:

Q = heat (Btu/hr) recovered by water
m = mass flow rate (lb/hr) of exhaust gas
c_p = heat capacity (Btu/lb°F) of exhaust gas
T_2 = exhaust gas temperature (°F) before heat recovery
T_1 = exhaust gas temperature (°F) after heat recovery

Heat Exchangers

Heat exchangers usually are of a tube and shell type. A bundle of tubes is arranged inside a shell, and hot fluid flows through the tubes while cold fluid passes outside the tubes. The exchanger can be a single- or multiple-stage device. The formula for the heat exchanger is the same as that for cooling towers:

$$Q = U \cdot A \cdot (LMTD)$$

where:

Q = heat (Btu/hr) exchanged
U = overall heat transfer coefficient (Btu/hr ft^2 °F)
A = heat transfer surface area (ft^2)

$(LMTD)$ = log mean temp. difference (°F)

$\quad\quad\quad\quad = [(T_{i2} - T_{o2}) - (T_{i1} - T_{o1})]/\ln{[(T_{i2} - T_{o2})/(T_{i1} - T_{o1})]}$

T_{i2} = inner hot fluid exit temp. (°F)

T_{o2} = outer cold water entrance temp. (°F)

T_{o1} = outer warm water exit temp. (°F)

T_{i1} = inner hot fluid entrance temp. (°F)

The overall heat transfer coefficient, U, has the following dimensions:

U	*Heat Transferred*	
(Btu/hr ft^2 °F)	*From*	*To*
250	hot vapor	boiling liquid
150	hot vapor	flowing liquid
50	liquid	liquid
20	gas	liquid
10	gas	gas

Gas-to-gas heat exchangers have been built for application in recuperators, gas turbine regenerators, and pipes and ducts. Gas-to-liquid heat exchangers are used in finned-tube economizers, tube-and-shell heat exchangers, and waste heat boilers. Liquid-to-liquid heat exchangers are used mainly in tube-and-shell heat exchangers.

Heat Recovery in Various Operations

Other possibilities for recovering waste heat exist in such operations as heating offices and homes, warm water irrigation, heating processes in salt-water desalination plants, and heating the building complexes of sewage treatment plants. Knowledge of such applications is essential to city planners who need to take waste heat recovery into consideration. The key consideration is a comparison of the waste energy recovery and the human health risk caused by the resultant air pollution.

ENVIRONMENTAL IMPACTS OF VARIOUS COOLING TECHNIQUES

The use of the aforementioned cooling techniques affects the immediate environment as discussed below.

Waste Heat Addition to a Natural Body of Water

When heat is added to water, there is a tendency for the following to occur:

- The water temperature increases, and the oxygen content decreases. Oxygen-deficient water will kill fish.
- Chemical reactions increase, so that the toxicity of the water increases. An increase in chemical reactions generally will produce a more hazardous environment for fish.
- Water layers of different temperatures are formed. Because fish depend on temperature as a signal for migration or spawning, a wrong signal will have harmful effects. Also, above a certain temperature eggs will not hatch, and changing the temperature of the water causes the bodily functions of fish to change.
- The viscosity of the water decreases. A change of water viscosity will change the habitat of fish and will affect their life span.

Therefore, using a once-through cooling system to remove waste heat is less advantageous than it might seem.

Cooling Ponds

A cooling pond requires a large piece of land that otherwise might be used more beneficially. Also, local weather can be affected by its evaporation. The advantages of a cooling pond are in conducting warm-water sports activities and in cultivating warm-water fish.

Cooling Towers

Aesthetically, a cooling tower is undesirable, and gaseous pollutants may be emitted to the atmosphere from the tower. In addition, a wet cooling tower will spread moisture, causing foggy and icy roads that can create dangerous traffic conditions.

Heat Transfer to Other Systems

For waste heat to be transferred to another system, heat sources must be installed near urban areas. The associated cancer risk due to human exposure to carcinogenic air contaminants or radioactive waste should be determined.

Waste heat management depends on many factors, and has no single solution. The final choice can be made only when all the factors are considered for a particular location under specific conditions, even though some uncertainty will remain.

A specific application for waste heat from spent steam has opened up a new area of energy conservation known as cogeneration.

COGENERATION

Cogeneration is defined as the production of electricity and thermal energy from the same fuel source. At first glance, the concept seems economically sound; however, it does have cost problems. This section discusses types of cogeneration operating cycles, control of air pollutants, and cogeneration problems.

Types of Cogeneration Operating Cycles

Two types of operating cycles are presently used: topping and bottoming cycles. A topping cycle begins with burning fuel to generate electricity and ends with industrial processes. A bottoming cycle begins with burning fuel to support industrial processes and ends with electricity generation. Examples of these cycles are discussed below.

Topping Cycles

Example 1: Steam Turbine Power Generation
Fuels are burned to produce steam in a boiler, and the steam drives a turbine coupled with a generator to generate electricity. The spent steam is used in processes of manufacturing and heating and in cooling equipment.

Example 2: Gas Turbine or Diesel Engine Power Generation
Fuels are burned to fire a gas turbine or a diesel engine to drive a generator, which generates electricity. The exhaust gases pass through a waste heat boiler to produce steam for process uses.

Bottoming Cycles
Fuels are consumed for manufacturing processes, and the waste heat from the exhaust gases is guided to a boiler in which supplemental fuels are added to produce steam for a turbine. A power generator is driven by the steam turbine to generate electricity.

All these operating cycles emit similar air pollutants, which need to be controlled properly.

Air Pollution Control

Emissions from the exhaust gases of a cogeneration project contribute to the concentration of pollutants in an already contaminated atmosphere. The quantity of emissions is far beyond the legally allowed threshold limits; so control of the emissions is required, and the most stringent control devices are mandated in most nonattainment areas.

The two major pollutants emitted from such projects are carbon monoxide (CO) and oxides of nitrogen (NO_x). Their control measures are the following: For CO emissions, a CO catalyst, a precious metal, is used to accelerate the oxidation of CO to CO_2. For NO_x emissions, a combination of water injection and Selective Catalytic Reduction (SCR) is considered to be the best available control technology. The water may be injected either into the combustion chamber or into the fuel oil.

The injection of water into the combustion chamber will suppress soot formation, so that the amount of excess air can be reduced in the combustion process. This helps to reduce thermal NO_x formation.

Mixing water with fuel oil separates oil droplets into very fine particles and increases the contact area of oil with air, thereby shortening the combustion time and reducing thermal NO_x formation. It has been reported that the NO_x concentration can be reduced to 24 ppm by water injection.

The NO_x concentration can be reduced further, from 24 ppm to 9 ppm, by using SCR, which is a proven method of NO_x control for gas turbines in both the United States and Japan. SCR involves injecting ammonia into an exhaust gas. The resultant mixture reacts with oxygen in the presence of a catalyst, converting the mixture into nitrogen and water:

$$4NO + 4NH_3 + O_2 \xrightarrow{\text{Catalyst}} 4N_2 + 6H_2O$$

$$2NO_2 + 4NH_3 + O_2 \xrightarrow{\text{Catalyst}} 3N_2 + 6H_2O$$

The amonia-to-NO_x ratio is approximately 1:1, and the catalyst can be $CuSO_4$ or TiO_2 (titania) plus 10 percent V_2O_5 (vanadium oxide).

Cogeneration Problems

The problems attributed to cogeneration projects are those of cost and time. The time required to build a cogeneration plant may last for years so that the total cost would be affected. The capital costs of associated equipment should include the costs of storage tanks for ammonia, catalysts, and wastes generated during the operation. The maintenance costs and hazardous waste disposal costs also are important considerations.

8

Spillage and Leakage

Spills and leaks of petroleum products generally occur as oil spills and underground storage tank leakage. Whether it is from accidental oil spills or the perpetual leakage of aged storage tanks, the environmental impacts of this pollution are enormous. Causes of and methods for preventing spills and leaks, as well as technologies relevant to remedial actions, are discussed below.

OIL SPILLS

Introduction

Oil spills are events in which oil is spread in an ocean, or comes in contact with water in an area where it is not supposed to occur. Thousands of tankers transport oil from one country to another, and vessels have dumped petroleum products into the oceans for many years; yet with oil constantly being transmitted around the world in various forms, spills on the high seas are not the major pollution source. Instead, the major problem is the average day-to-day oil spills that occur in harbors, rivers, lakes, and streams, approximately two-thirds of which are due to human error. Whether the captain of a barge has fallen asleep during an oil transfer operation, causing a tank to overflow, or a plant operator has forgotten to regulate a valve at a refinery plant, causing a sprung valve and oil seepage into a river, or a maintenance worker has ignored a service schedule, causing a hose to burst and oil to erupt at a facility—the oil spill problem remains the same.

Oil spills also occur because of collisions of tankers or barges, oil well blowouts, leakage of oil tanks or pipelines, or the dumping of used oil. The following list shows the approximate sources of these spills:

- 22 percent from tanker operation.
- 3 percent from tanker accidents.
- 10 percent from normal sea transportation.
- 1 percent from offshore oil production.
- 13 percent from coastal facilities.
- 41 percent from river and urban runoff.
- 10 percent from natural seepage and other causes.

Spilled oil disperses into its surroundings, initially spreading widely over the water and then evaporating into the atmosphere.

Oil Spill Dispersion

Oil spills behave as follows: Initially, the spilled oil sticks together. Then, because of water currents and the wind, it tends to spread out. Warm weather increases the spreading. After the oil has spread, volatile compounds are evaporated by wind, solar radiation, and wave action. Generally the most toxic compounds are the most volatile; the remaining, less volatile compounds are split into small droplets, which form a thin film known as an oil slick.

The oil slick absorbs ultraviolet light from the sun, and the oil molecules become activated. They react rapidly with oxygen, which converts hydrocarbons (HC) into hydrogen peroxide (H_2O_2), phenol (C_6H_5OH), and aromatic compounds. These products, which are more soluble and more toxic than the original compounds, disappear quickly from the surface, but they have a more negative effect on the environment than does the original spill. The remaining portion of the oil floating on the surface either reacts with surrounding bacteria or adheres to sediments, such as clays, and precipitates to the bottom. It covers the surface of aquatic organisms or other sediment.

After the precipitated oil reaches the bottom, the oil begins to migrate laterally from one sediment to another, affecting a wide range of biotic communities. It comes to rest after a long period of time in deeper and softer deposits. When the oil reaches a soft deposit, it may cease to migrate laterally but gradually may float back from the sediment into the overlying water column to become a chronic source of water pollution.

Environmental Impact of Oil Spills

Spilled oil affects air, water, and land by its evaporation, spread, and migration. Its impacts are as follows:

- The toxic compounds that evaporate are transported inland by the wind and affect the air quality of nearby communities.
- When an oil slick covers fish eggs, the eggs cannot hatch. Large amounts of codfish and pollack are lost because of oil slicks covering their eggs.
- Marine birds and coastal birds are susceptible to oil spills. Sea birds frequently seek oil slicks for diving, apparently because fish gather there.
- Some hydrocarbons are soluble in water, especially the aromatic hydrocarbons, which are very toxic, causing great environmental damage.
- Underwater, or benthic communities, are damaged.
- Wetlands are destroyed by spilled oil.

Prevention of Oil Spills

The most effective way to avoid the damage due to oil spills is to prevent their occurrence. Oil spills are due primarily to careless personnel or inadequate, aged equipment; so the preventive measures, listed below, include careful selection of personnel and diligent equipment maintenance:

- Carefully select responsible and well-trained personnel.
- Regularly check transportation vehicles.
- Adequately design transportation vehicles.
- Properly equip transportation vehicles with chemicals to absorb spilled oil or to burn it safely.
- Have available devices that can remove oil slicks.

A device commonly used to remove oil slicks is known as an oil boom. It consists of cylindrical float sections, made of either polyethylene or polyurethane, and vertical plastic curtains. Each plastic curtain is attached to a float section that is bound to a cable. The boom is distributed around the spilled oil, and skimmers, or suction heads, are used to remove the oil for recovery.

In designing an oil boom, several parameters are important; it should be able to float, to follow wave contours, to withstand natural forces, and to maintain its physical integrity when controlling oil spills. Also, it should be easy to handle, easy to store, and easy to apply under actual spill conditions.

Oily Waste Treatment

Oily Waste Processing Plants

The oily fluid accumulated from spilled oil or any other oily waste can be treated and brought to a recoverable state. A typical oily waste processing plant consists of an oily waste collection pit, an oil–water gravity separator, a holding tank, an oil–water separator filtration system, a clean oil storage tank, and a pH adjustment tank.

Oily waste is collected in the collection pit, where a suction header is placed near the pit bottom to transfer the waste to a gravity separator. The gravity separator contains a filter, vertical baffles, and inclined parallel settler plates. As oily waste passes through the filter, coarse solids are removed. Other solids are precipitated when the waste passes through a series of baffles and inclined settlers. Oil and water are separated by gravity.

In the upper portion of the separator, oil skimmers are connected to oil lines that suck and transfer the upper-layer oil to the holding tank. Sulfuric acid may be used to break up emulsions in the holding tank, allowing the bound oil to separate from the water. The upper-layer oil is pumped from the holding tank to the filtration system, where lime is added to convert metallic substances to salts and precipitate them. Once the water and the metallic substances are removed, the clean oil is pumped to a storage tank. The solid residues are hauled away for recovery, and the wastewater is guided away for treatment to meet specific standards for discharge before being discharged to the sewer system.

The most common measure of water quality is the pH value, the acidity; the allowable pH range is between 5 and 11. By adding hydrochloric acid or sulfuric acid, the pH can be adjusted from its upper end (alkali) toward its allowable limit (neutral); by adding sodium hydroxide or sodium bisulfate, the pH can be adjusted from its lower end toward the neutral range.

The standards for substances contained in sewerage differ from one region of the United States to another. The Industrial Waste Program Guidance for Industrial Users, Metropolitan Sewerage System, San Diego, California specifies the following specific discharge limitations:

pH	5.0–11.0
Arsenic	2.0 mg/L
Beryllium	2.0 mg/L
Cadmium	1.2 mg/L
Chromium	7.0 mg/L
Copper	4.5 mg/L
Cyanide	1.9 mg/L
Grease and oil	500.0 mg/L
Lead	0.6 mg/L
Mercury	2.0 mg/L
Nickel	4.1 mg/L
Pesticides and PCB	0.04 mg/L
Phenolic compounds	25.0 mg/L
Selenium	2.0 mg/L
Silver	2.0 mg/L
Sulfides	1.0 mg/L
Zinc	4.2 mg/L

A modern oily waste processing plant must comply with the following requirements (California Health & Safety Code, Section 25280, *et seq.*):

- All below-ground level waste collection sumps shall be double-contained. In an old plant, a collection tank can be inserted inside the existing sump.
- All pressurized waste oil transfer lines shall be double-contained.
- Any in-ground sumps that can be operated aboveground shall be removed and replaced by an aboveground sump with secondary containments.
- All above-ground treatment equipment shall be placed on a concrete slab, surrounded by a wall with a secondary containment and an alarm monitoring system.

Summary of Oil Recovery Technologies
The modern technologies used to recover waste oil are summarized below.

Gravity Differential Separation
Because of the different specific weights of individual compounds, waste oil is differentiated into different layers by the force of gravity. Gravity separates water and solids from oil, making possible the recovery of free oil.

Vacuum Filtration
A filter is set on a vacuum rotary drum, and when waste oil is loaded into the filter, liquid is spun out through it and sucked away by a vacuum. Solids remain inside the filter and are removed from the system with the filter.

Acid Treatment
Soap or detergent is used to remove oils. The resulting mixture of oil, soaps, and water becomes highly alkaline, an emulsified waste oil. Sulfuric acid or hydrochloric acid treatment can break the emulsion and separate out the saturated naphthenic and paraffinic molecules. Free oil can be removed for further refining.

Electrostatic Cleaning
Electrodes are electrostatically charged, and when waste oil is placed near them, they remove sediment from the waste. The remaining liquid is free of solids.

Chemical Treatment
A dissolved metal, such as a zinc (Zn) or other salt of polyvalent metals, is used to salt out the alkali soaps:

$$Zn^{+2} + 2\,(OH)^- \rightarrow Zn(OH)_2$$

and to destroy the emulsifying agents. The addition of flocculants, which are fine solids, accelerates salt removal.

Flocculation and Sedimentation
When finely divided solids, such as clay or fly ash, are added, the salts are aggregated (flocculated) and can be removed faster. Even if wastewater sludges are added to waste oil, emulsions are separated because the mixed liquid becomes a semisolid (coagulates).

Agitation
Emulsions may be broken simply by vigorous agitation, which usually is carried out by motor-driven blades in a reactor. Agitation, then, is a mechanical method used to separate oil from a water mixture. The separated oil can be treated further.

Ultrasonic Vibration
Ultrasonic vibration is the use of high-frequency sound vibration to break emulsions. Sound waves are transmitted to oil and water particles, which have different densities, causing each particle to vibrate at a different amplitude and phase, so that the oil particles are separated from the water effluent. The separated oil is removed for further recovery.

Ultimate Disposal

Land Disposal
Land disposal of oily waste is accomplished by mixing the oily sludge with soil, using a bulldozer. It usually requires several acres of land; oily sludge and soil are spread 6 inches deep, two to four times per month, on the land until the hydrocarbons have been consumed by bacteria. This method cannot be used, however, because it results in underground water contamination.

Incineration of Oily Substances
Oily substances can be incinerated either by a fluidized bed or by a rotary kiln. Unless the quantity of waste oil is very great, the capital cost of this method is high and economically prohibitive.

UNDERGROUND STORAGE TANK LEAKAGE

Introduction

An underground storage tank is a tank or a combination of tanks used to contain regulated substances that has 10 percent or more of its volume beneath the surface of the ground. The underground piping connected to such a tank is included in volume considerations. The regulated substances are crude oil, petroleum products, or any hazardous substance as defined by CERCLA (the Comprehensive Environmental Response, Compensation, and Liability Act).

Statistical data indicate that three to five million underground storage tanks now exist in the United States. It is estimated that 100,000 tanks are leaking, and that approximately 300,000 tanks will be leaking by 1997. This leakage will contaminate the soil and the groundwater, it will threaten the health and safety of humans and wildlife, and it will pose a danger of explosion because it is confined leakage.

Lawmakers have responded to the leakage at both federal and state levels. The Resource Conservation and Recovery Act and the Hazardous and Solid Waste Amendments of 1984 both contained mandates for the regulation of underground storage tanks; and with the passage of the Superfund Amendment and Reauthorization Act, additional funds were provided for underground storage tank control. On the state level, the California Health and Safety Code specifies standards for underground storage tanks for hazardous substances. Also the California Administrative Code indicates how to manage and report the hazardous substances in underground storage tanks. The California Health and Safety Code, Section 25284, specifies that a person cannot own or operate an underground storage tank unless a permit has been applied for and obtained. (Businesses or individuals acquiring real estate should be aware that hidden underground tanks are subject to regulation.)

The state of California requires monitoring for oil tanks that were installed on or before January 1, 1984. The monitoring system must be capable of determining the containment ability of the underground storage tank and detecting any active or future unauthorized releases. If a storage tank was installed after January 1, 1984, secondary containment and monitoring systems are required.

It is necessary to report to the local air quality control agency within 24 hours of the occurrence of any leaks or other releases if there is no secondary containment. If secondary containment has prevented a release to the atmosphere and the release has been cleaned up within 8 hours, no report is necessary.

Basic Underground Storage Tank System

The basic underground storage tank system consists of three subsystems: tanks, piping, and accessories. The tanks include primary and secondary tanks, mainly for containment of the product. The piping system conveys and transfers the product from one point to another within the system. Accessories include pumps, valves, vapor vents, and vapor return lines, which control and regulate the flow of the product and the operation of the system. The tanks, the piping system, and the accessories all are included in one permit.

Causes of Tank Leakage

Old tanks are constructed of steel and are not protected against rust. When these tanks are buried underground, the surrounding soil contributes to electrochemical activity, causing corrosion of the external steel surface. Corrosion generally occurs because of:

- Improper tank installation or handling, causing structural failure.
- Incompatibility of the tank's contents with its construction or liner materials.
- Poor operating practices.

Corrosion involves the loss of electrons from metals to other elements (oxidation) to form metal ions. For example, when un-ionized iron loses two electrons, Fe^{+2} is formed:

$$Fe^0 \rightarrow Fe^{+2} + 2e$$

Corrosion can be prevented by stabilizing the current charge (adding a cathodic corrosion protection layer such as copper) or by adding some agent (such as silicate, which forms sodium silicate as a barrier) to disrupt the electrical connection. Thus stabilization of the current charge can be accomplished as follows:

$$Cu^{+2} + 2e \rightarrow Cu^0$$

In this (reduction) reaction electrons are gained. Thus oxidation (loss of electrons) causes corrosion, and reduction (gain of electrons) prevents it.

Leak Detection Methods

There are internal and external means of detecting leakage of underground storage tanks. Internal methods include the following:

- *Inventory balance:* The amount of material purchased should balance the amount used and the inventory; otherwise, leakage has occurred.
- *Liquid level monitoring:* The overnight liquid level should stay constant.
- *Pressure testing:* When liquid leaks out of a tank, the pressure inside the tank decreases.
- *Acoustical monitoring:* Sound waves should be sent into the system and a response obtained. When leakages exist, the response frequency varies.

The external leak detection methods include:

- *Monitoring between primary and secondary containments:* When a leakage occurs, an alarm is triggered.

- *Groundwater table monitoring:* Samples are taken from the groundwater and analyzed.
- *Soil vapor monitoring:* Perforated tubes are inserted around the site and the extracted vapors are checked. Different intensities of vapor pressure indicate the different degrees of soil contamination.
- *U-tube monitoring:* A U-tube manometer is installed with one side connected to the upper edge of the tank and the other side to the atmosphere. When leakage occurs, the manometer reading varies.
- *Soil core testing:* A sample is taken from the surrounding soil for laboratory analysis. Chemical compounds can be determined explicitly.
- *Surface geophysical monitoring:* A surface topographical survey is performed. When contaminated soil erodes, the physical properties of the soil change.

After leakages are detected, immediate correction is required. The following are some remediation measures.

Remedial Actions

A number of remedial measures are employed, as described below.

Preventive Measures
- Proper equipment installation can prevent damage and reduce corrosion. For stainless steel, reinforced plastic containers can prevent corrosion.
- The primary and secondary containments should be designed properly. The secondary containment can be concrete walls or a flexible membrane.
- Installation of a cathodic corrosion protection layer will minimize corrosion.
- Selecting compatible liquids and container materials will prevent unexpected corrosion.

Plume Management
Plume management is the effort to isolate contaminated soil and surface water from a specific site. Various control measures can be used:

- Contaminated soil is excavated and removed. A slurry wall, grout curtains, and sheet covers can be installed as physical barriers.
- Contaminated surface water is controlled by using vegetation, channels, and vertical pipes to keep the contaminants from spreading, and by directing the water through a bypass leading away from the groundwater.
- A passive means of control is to provide ditches alongside the underground storage tank to collect the leaking liquid and remove it for treatment.
- Active ways of controlling contaminated surface water include the installation of a leak detector in the secondary containment, laying a cathodic corrosion

protector beneath the primary tank, and providing a monitor well with a liquid detector.
- Aged tanks with major leakage must be removed and disposed of.

Tank Removal and Disposal

Before a tank is removed, all the liquid must be transferred to another storage tank. The vapor remaining in the tank must be pumped either to a carbon adsorber or to an internal combustion engine to destroy it. Upon the excavation of the tank, if the soil is contaminated, a plastic cover sheet or a foam spray is required to depress volatile organic compound (VOC) evaporation to the atmosphere. Excessively contaminated excavated soil must be treated on site by either thermal oxidation or biological methods (Chapter 5). Another alternative is to confine the contaminated soil in sealed containers and to move them to designated facilities.

Factors that must be considered before removing an underground storage tank are:

- The soil's resistivity.
- Low soil pH values.
- The groundwater height near the tank.
- Any new tank replacement.
- The building foundation near the tank.
- The age of the tank.
- The cost of its disposal.

The tank may be removed and disposed of after the safety and cost factors are taken into consideration.

9

Petroleum Refining Operation

Petroleum has played an important role as an energy source for decades. Following brief introductory sections, this chapter discusses the most essential petroleum refining processes, including treatment methods, which are grouped into primary and secondary operations.

INTRODUCTION

Crude oil contains water, salt, and sand, which must be removed before the oil refining processes can begin. Some water and sand settle out during storage, but other saturated water and dissolved salts must be removed through separation processes. After separation, crude oil becomes petroleum gas, naphtha, gas oil, and coke. The larger hydrocarbon chains have to be broken into smaller chains through decomposition processes. For example, gas oil is decomposed into gasoline or distillate fuel. The quality and the octane number of these compounds can be improved through a reforming process. The octane number, which indicates the antiknock properties of an automobile fuel mixture under standard test conditions, is zero when the engine performance is equivalent to that of pure normal heptane, a very high-knock fuel; it is 100 when the degree of knocking is as low as that of isooctane. Impurities and harmful compounds in the products can be removed through treatment processes; the released gases are recovered before entering the atmosphere. The chemical composition of petroleum is summarized below.

Composition of Petroleum

Petroleum, also known as crude oil, is a complex mixture of hydrocarbons and nonhydrocarbon compounds. Many products are manufactured from crude oil, either by rearranging the hydrocarbons in the feedstock or by breaking down complex hydrocarbons into simpler ones. Depending upon the number of carbon atoms, petroleum products reveal themselves in different phases at normal temperature and atmospheric pressure. The products are gaseous when their molecules contain 4 or fewer carbon atoms; they are solid when the carbon atoms number 20 or more. Liquid petroleum may have between 4 and 20 carbon atoms per molecule; the liquid mixture may contain gaseous and/or solid compounds in solution.

Each carbon atom has four bonding sites, which may be linked to hydrogen atoms, to other atoms or compounds, or to a nearby carbon atom (by single, double, or triple bonds). Hydrocarbons are said to be saturated when the full complement of hydrogen is attached to all carbon bonds; otherwise they are unsaturated hydrocarbons (contain double or triple bonds). Unsaturated hydrocarbons are more reactive with other compounds than saturated hydrocarbons are.

The arrangement of hydrocarbons creates three basic structures: straight-chain, branched-chain, and ring-type compounds, which are known as paraffins, isoparaffins, and cycloparaffins or naphthenes, respectively. The paraffins (straight-chain hydrocarbons), such as butane (C_4H_{10}), *n*-hexane (C_6H_{14}), naphtha, or wax, have very low octane numbers. The isoparaffins (branched-chain hydrocarbons), such as isobutane and isopentane, have relatively high octane numbers. The naphthenes (cylcoparaffins) have different antiknock characteristics, depending upon the structure of the ring compounds. Generally, ring compounds containing one or more six-membered rings with three alternate double bonds (aromatic groups) have higher antiknock properties than more saturated ring compounds. Compounds with low octane numbers are further processed to high-antiknock products.

Straight- or branched-chain hydrocarbons with one double bond (ethylene, C_2H_4; mesityl oxide, $C_6H_{10}O$) are known as mono-olefins (or alkenes); those with two double bonds in the structure (butadienes, C_4H_6) as diolefines (or dienes); those with a triple bond (acetylene, C_2H_2) as alkynes. Usually crude oil does not contain olefins; they are obtained mainly through synthetic methods, by converting long-chain hydrocarbons into antiknock additives.

In comparison with hydrocarbons, the quantity of nonhydrocarbons contained in crude oil is relatively small; yet their effects on knocking are enormous. The major nonhydrocarbon compounds in crude oil are sulfur, nitrogen, oxygen, and chlorine. Their essential effects are described in the following paragraphs.

Sulfur compounds appear in crude oil processes in different forms: free sulfur, hydrogen sulfide, ethanethiol (C_2H_5SH), diethyl sulfide ($C_2H_5SC_2H_5$), diethyl disulfide ($C_2H_5S_2C_2H_5$), and so on. These compounds are formed in distillation processes. Although some of these compounds are noncorrosive, at high temperatures a noncorrosive compound such as diethyl sulfide is decomposed to form a corrosive compound such as ethanethiol that corrodes process equipment.

Sulfur compounds in fuels corrode equipment, in addition to having a knocking effect in engines. The combustion products yielded by the combustion of fuels that contain sulfur are sulfur dioxide (SO_2) and sulfur trioxide (SO_3), which form sulfuric acid when exposed to water vapor or water. When deposited on surfaces, the sulfuric acid causes white spots (on lamp glass), bad odors, and discoloration (on painted surfaces). Therefore, the sulfur compounds in fuels should be removed as thoroughly as possible.

Unlike sulfur compounds, most of the nitrogen compounds in crude oil are not identified; but during distillation processes, products containing a six-membered nitrogen-containing ring (pyridine) have been reported. In gasoline, nitrogen compounds cause discoloration, engine fouling, catalyst poisoning, and poor lubrication.

Few of the oxygen compounds contained in crude oil have been identified. Among the compounds found during distillation processes is naphthenic acid, a carboxylic acid group (COOH)–containing hydrocarbon ($C_7H_{13}COOH$). Other compounds formed during cracking processes include phenolic compounds (such as phenol), derived from the oxidation of aromatic compounds containing one or more hydroxide (OH) groups. Naphthenic acids are highly corrosive, and phenol is a toxic substance.

Chlorine compounds are found in crude oil in the form of chloride salts, such as sodium chloride (NaCl). The chloride salts are deposited on heat-transfer surfaces and decompose to form acids, causing equipment fouling, corrosion, and catalyst poisoning.

Crude oil also contains metallic elements, such as vanadium, nickel, sodium, potassium, copper, zinc, and iron. These elements are found in salt form, and most of them will spoil catalyst activity.

All of these nonhydrocarbon compounds are either formed by or released from different refining processes. They are treated during the processes, or they are controlled after they are formed but before being released into the atmosphere. Major processes in the petroleum refining operation are discussed below.

PRIMARY OPERATIONS

The primary operations in petroleum refining industries mainly are processes for separation, decomposition, and formation. Each process can be further broken down into subprocess units.

Separation Processes

Separation processes include desalting, distillation, and deasphalting units.

Desalting Unit

A desalting unit can be an electrical or a chemical desalting vessel, where water and crude oil are mixed. Salts are dissolved in the water, and the water and oil phases can be separated in a settling tank.

For electrical desalting, an electrostatic field is created by introducing approximately 15 to 35 kilovolts of electricity to the emulsified mixture at a temperature between 150°F and 300°F. The water pressure should be high enough to prevent vaporization of the water. The impurities in the mixture will grow in size to form ball-shaped substances that can be removed from the solution.

For chemical desalting, a chemical such as hydroxyacetic acid ($CH_2OHCOOH$) is added to the water–crude oil mixture to break up emulsions. The water phase with its impurities is settled and removed from the crude oil. The water temperature is approximately 250°F to 300°F, and the amount of water to be added is approximately 3 to 10 percent by volume, depending upon the density of the crude oil.

After the water phase and the impurities are removed, the desalted crude oil is ready to be fed to distillation units.

Distillation Unit

A distillation unit contains distillation towers, stripping towers, overhead accumulators, crude oil heaters, heat exchangers, condensers, blowers, and pumps. The main purpose of a distillation unit is to use steam as a heat source to separate crude oil into different products based on different boiling points. Its overall functions include heating, vaporizing, condensing, fractionating, and cooling. The products are, in descending order of quality: petroleum gases, light naphtha, heavy naphtha, light gas oil, and heavy gas oil. At the end of the distillation process, the residue is further handled in a deasphalting unit (discussed below). A distillation process is completed (under atmospheric or vacuum conditions) in two steps: evaporation and condensation.

Atmospheric distillation is carried out mainly in a vertical, atmospheric distillation column. Several trays are installed in the column to cover most of a cross-sectional area, except for channels carrying liquid from the upper trays down to the lower trays. The upper portions of the channels form weirs for individual trays, and the lower portions act as downcomers. Thus each section of the column contains a weir, a downcomer, and a perforated tray.

Before entering the distillation column, crude oil first is preheated by a heat exchanger and a crude oil heater. The crude oil temperature is increased to approximately 750°F. The hot crude oil forms a vapor/liquid mixture and enters the distillation column from the lower section, as steam is continuously blown into the bottom portion of the column. The vapor and the steam pass up through the column, and gas, gasoline, and steam are removed from the top of the column to an overhead accumulator. The gasoline and the steam are condensed; the noncondensable gas is separated from the gasoline to feed the gas system where petroleum gas is compressed into liquid fuel. Water is removed from the gasoline by settling, and one part of the gasoline is returned to the column to maintain a downward-flowing stream of liquid, known as reflux.

When the crude oil and refluxing mixture spreads on a tray, it becomes a froth because of the gas and steam bubbling and jetting through the perforated tray. The froth is separated into two parts: vapor and liquid. The liquid flows over the weir and through the downcomer down to the next tray. The vapor travels up and through the upper perforated tray and mixes with the froth on the upper tray. The vapor is condensed, its latent heat is released, and hydrocarbons with lower boiling points are evaporated up to the next tray froth.

Some distillation products are removed from the sidestream of the column, which is a section between two consecutive trays. In order to improve the quality of the fractionation products, each sidestream is sent through a stripper, a small fractionating column in which all hydrocarbons with lower boiling points than that required of the products are evaporated and sent back to the main column. The heat source for the evaporation is steam injection into the bottom of the stripper. The bottom products are cooled and sent to storage.

In addition to the petroleum gas and gasoline, heavy naphtha, light gas oil, and heavy gas oil are among the most common sidestream products of atmospheric distillation columns. The residual oil accumulated at the bottom of the column, known as the long residue, can be further fractionated in a vacuum distillation tower.

A vacuum distillation tower is used after atmospheric distillation for further distillation of products with a smaller energy supply. It is a vertical tower with a steam ejector installed at the uppermost portion of the tower to maintain approximately a 40 mm Hg vacuum in the tower. Instead of simple trays, packed sections are preferred for vacuum distillation. A packed section is a pile of small rings, approximately 2 inches in diameter, with extended contact surfaces inside the rings, resting on a support grid. The height of the section can be 3 feet or more. A liquid distributor or a distributor–tray–distributor combination is installed atop a packed section. When the liquid is sprayed over the packing, films are formed on the contact surfaces. When hot gas is passed upward through the packing, evaporation and condensation of the liquid films occur.

The preheated (approximately 750°F) long residue and steam are injected into the lower portion of the tower. After passing through the packing, which reduces the levels of dirty spots, coke, and stained products, the sidestreams can be extracted from the trays to accumulators. Such products as light gas oil, a waxy distillate, and heavy gas oil are obtained from the different sidestreams. One portion of the product is returned to the distributors as a reflux. The residue obtained from the bottom of the vacuum distillation tower, called the short residue, can be further separated in a deasphalting unit.

Deasphalting Unit

The short residue contains asphaltenic and oily compounds, which can be separated by liquid–liquid extraction. This process uses an extraction liquid such as propane, butane, or pentane to extract other liquids or oil, mostly paraffins. It normally is carried out in a rotating disk contactor, which is a vertical column consisting of rotating and stationary parts. The rotating part is a motor-driven central shaft with disks as a rotor. The stationary part is the column separated by horizontal plates (the stator) installed between rotating disks. The space between the rotor and the stator allows the extraction liquid and the short residue to flow freely.

The short residue enters the contactor at the top, and the extraction liquid (propane) enters at the bottom. After the residue and propane are blended, the paraffins are dissolved by propane, forming oil and asphalt phases. Then they are gravitationally separated from each other, because of their different densities inside the contactor. The oil phase is removed from the top, and the asphalt phase from the bottom of the contactor. Propanes contained in both phases are steam-stripped, condensed, and recycled. The products are high-quality lubricating oil and asphalt. Butane and pentane also can be used as extraction liquids; they will yield more lubricating oil, but it will be of a lower quality than those products obtained by using propane as an extraction liquid.

After the separation processes, crude oil is fractionated into various products, each with its own hydrocarbon number and structure. The heavier hydrocarbons can be decomposed, forming lighter hydrocarbon compounds that are lower-boiling and more volatile than the heavier hydrocarbons.

Decomposition Processes

The processes of converting heavy hydrocarbon molecules into lighter molecules are known as decomposition processes. They include thermal cracking, catalytic cracking, and hydrocracking units. Each unit has unique applications.

Thermal Cracking Units

Thermal cracking units are used to convert (crack) long-chain hydrocarbons at a high temperature, approximately 900°F. The feed usually is the residue from atmospheric or vacuum distillation units. Because of insufficient hydrogen, after cracking, some carbon atoms cannot be saturated. Then the carbon bond temporarily joins with another carbon bond, forming an olefin, an unsaturated compound.

The products of thermal cracking are petroleum gases, light hydrocarbons (in the range between gasoline and oil), and heavier products. The disadvantages of this method are its high fuel consumption, inefficiency, and sludge and tar residues. Therefore, for many purposes, such as producing gasoline or light hydrocarbons, thermal cracking is replaced by catalytic cracking. This technology is explained below (under "Catalytic Cracking Unit") after a discussion of the main applications of thermal cracking: visbreaking and coking systems.

Visbreaking Systems
Visbreaking systems are used to reduce the viscosity of short residues so that smaller amounts of diluents will be required for their conversion into fuel oils. The system includes a furnace, a reaction chamber, a fractionator, a knockout pot, and steam strippers. The short residue is heated in the furnace to approximately 860°F and then sent to a reaction chamber, where the residence time can be extended as necessary. The products' temperature at the outlet of the chamber is approximately 800°F; the pressure inside the chamber is between 5 and 10 atmospheres. The products are cooled by recycled distillate to between 550°F and 750°F before entering the fractionator. The pressure inside the fractionator is approximately 2 to 5 atmospheres. Following the fractionator, the effluent passes through a knockout pot and steam strippers; and the products are petroleum gases, gasoline, gas oil, and a residue.

Coking Systems
Coking systems are used to drive off hydrocarbon vapors from heavy residues to increase the distillation products and to obtain coke as a residue. This is a severe application of thermal cracking; the system includes a furnace, two coke drums, and a fractionator. The operation usually is performed batchwise. For a continuous operation, the two drums should alternate, with one used in the operation and the other as a standby.

The heavy residue is heated to between 840°F and 930°F and then proceeds to a coke drum, where the operating pressure is 2 to 3 atmospheres. Vapors are driven off the residue to the top of the drum and are transported to the fractionator for further distillation, where the products are gas, gasoline, and gas oil. The residue remaining in the drum is coke.

This process is also known as delay coking because the coking operation occurs at the drum and after the furnace, instead of at the furnace.

Catalytic Cracking Units
Catalytic cracking units are much preferred over thermal cracking to obtain a high-octane petroleum product. Heavy fuel oil is cracked with the aid of catalysts to form olefins, double carbon bonds, and iso compounds, which are branched chains with a much higher octane number than that of the heavy fuel oil. The process yields include petroleum gases such as ethane, propane, and butane; the

major emissions are H_2S, NO_x, CO, particulate matters (PM), and small amounts of H_2. The flue gases are transferred to CO boilers to generate steam, which takes care of the CO emissions; electrostatic precipitators are used to control PM; caustic scrubbers are used to reduce H_2S and NO_x emissions.

During the cracking processes, coke is formed and deposited on catalysts, thereby hindering or deactivating catalytic reactions; so, for economic reasons, the spent catalysts need to be reactivated for reuse. The processes for recycling catalysts include cracking, stripping, and regeneration. In the cracking stage, residues fed to the process are converted to light hydrocarbons; in the stripping stage, steam is injected to remove hydrocarbons adhering to the catalysts; and the catalysts are regenerated by blowing air through them to burn the deposited coke. The regenerated catalysts are ready for reuse.

Depending upon the application of the catalysts, the cracking operations can be categorized as a fixed-bed catalyst cracking (Houdry) process, a moving-bed catalyst cracking (Thermofor) process, and a fluidized-bed catalyst cracking (FCC) process.

Fixed-Bed Catalyst Cracking (Houdry) Process
Catalysts originally were made of aluminum chloride. Later, because of its high cost and difficult reclamation, naturally occurring clays came into use. The size of the catalyst is approximately 0.25 inch in diameter in a pellet form. Pellet catalysts are set on a fixed bed contained in a vessel, where cracking, stripping, and regeneration operations are conducted sequentially. Generally three vessels form a complete operating set, to be used sequentially, one in the cracking operation, one in stripping, and one in regeneration. The duration of a complete operating cycle is less than one hour. This is a batch operation and is labor-intensive. A moving bed is used for continuous operation.

Moving-Bed Catalytic Cracking (Thermofor) Process
For a moving-bed catalytic cracking process, the reactor and the regenerator are separated but placed adjacent to one another. The spent catalysts are transported from the bottom of the reactor to the top of the regenerator, and the regenerated catalysts from the bottom of the regenerator to the top of the reactor. The reaction and the regeneration are performed continuously, and the catalysts are transferred by means of high-velocity gas flow. The temperature of the hot feed is approximately 750°F prior to contact with the pellet catalysts.

Fluidized-Bed Catalytic Cracking (FCC) Process
Modern catalytic cracking uses a fluidized bed, in a process known as fluid catalytic cracking. The catalysts are made of silica alumina or of crystalline materials known as zeolite catalysts. They are used in powder form with spheres measuring between 1 and 50 microns in diameter.

The powder catalysts are loaded into a reactor connected to a nearby regenerator, a riser, and two standpipes. The lower end of the riser is connected to the feed, and its upper end is attached to the reactor. The lower part of the reactor is connected to the regenerator by a standpipe sloping down from the reactor to the regenerator. Another standpipe joins the regenerator to the lower part of the riser.

When the hot feed and steam are blown at a rate of 3 to 20 inches/sec from the bottom of the riser upward through the powdered catalytic bed, the catalyst and the gas form a mixture that behaves like a liquid (is fluidized). Before reaching the top of the riser, the hot feed is cracked by the catalyst mixture and steam; the mixture thus is separated as it enters the reactor, and the hydrocarbon vapor fluidizes the catalyst.

The vapor passes on through a cyclone inside the reactor to drop the entrained catalytic particles, and the cleaned vapor flows to a distillation column for further fractionation. The spent catalyst is accumulated at the bottom of the reactor, after steam is injected to strip hydrocarbons from the catalytic powder. The spent catalyst powder is transferred via the standpipe down to the regenerator. A stream of air is blown to the regenerator to burn off coke, and regenerated catalyst is recycled back to its initial place on the riser to mix with the hot feed and steam, thus starting the cycle anew.

The operating temperature in the reactor is from 1100°F to 1200°F, and at the regenerator approximately 1300°F to 1400°F. Typical yields of (vacuum) distillate include petroleum gases, gasoline, light gas oil, heavy gas oil, and a small percentage of coke. Because of the high operating temperatures, CO emissions are converted to CO_2, so that in modern catalytic cracking units a waste heat boiler often is installed to recover thermal energy, rather than a CO boiler to convert CO to CO_2.

Hydrocracking Units
Hydrocracking is a combination of hydrogenation and catalytic cracking, its purpose being to convert heavy fuel oil to light fuel oil. This process is flexible enough to produce different grades of high-quality products. No coke by-product remains, and the entire feedstock can be converted into the required products. The conversion involves three steps: saturation of organic compounds, removal of impurities, and cracking of heavy hydrocarbon molecules. Usually two reactors are used.

The feedstock ranges from aromatic liquid products obtained from atmospheric distillation to heavy fractions of oil, including the vacuum residue. Hydrogen is injected into the preheated feed, and the mixture passes through a heater so that the aromatic and olefinic compounds are saturated by the injected hydrogen (hydrogenated). The saturated mixture is transferred to the first reactor with a fixed- or fluidized-bed catalyst. The catalysts in the reactor are sulfided metals, such as cobalt, molybdenum, and nickel. These materials bring about the reactions that hydrogenate the impurities, catalyzing the following conversions:

• Sulfur to hydrogen sulfide (desulfurization).
• Nitrogen to ammonia (denitrogenation).
• Oxygen to water vapor (deoxygenation).

The reaction temperature is approximately 750°F, at 200 to 300 atmospheres. With improvement of the catalyst (by using zeolite and rare-earth metals), it has been reported that the reaction pressure drops to between 70 and 150 atmospheres.

The reactor effluent is cooled and sent to a separator, where gases and liquid are separated. The gaseous compounds are treated by passing them through a caustic scrubber and are routed to the second reactor. The catalyst in the second reactor (for example, silica-alumina sulfide) is acidic. Hydrogen is introduced again, in the second reactor, causing a hydrocracking reaction. The cracked mixture is sent from the second reactor to the separator, where the gaseous compounds are recycled back to the reactor, and the liquid compounds are guided to an additional separator for ammonia removal. The remaining liquid is fed to a fractionator to produce liquefied petroleum gas (gas oil). The emissions from the fractionator can be controlled by the caustic scrubber.

All the cracking processes described above involve breaking down heavy hydrocarbon molecules into lightweight molecules for high-octane performance characteristics. Another approach is to form desired molecular structures with the aid of different catalysts on different feeds, as discussed below.

Formation Processes

The formation of high-octane hydrocarbons containing a benzene ring, a branched chain, or a macromolecule can be briefly subdivided into the following processes: catalytic reformation, alkylation, isomerization, and polymerization. The individual procedures are described below.

Catalytic Reformation
Catalytic reformation, also known as dehydro-cyclization, is a process used to move hydrocarbon away from low-octane heavy gasoline fractions, such as naphtha and hexane, to form liquefied petroleum gas and reformate, for gasoline blending purposes. The equipment needed for these purposes mainly consists of heaters, separators, fractionators, and compressors.

Low-octane, desulfurized feed is heated at a heater before entering a reactor, where a section of catalyst rests on a fixed bed. The most commonly used catalyst materials are platinum and halogen dispersed on aluminum oxide carriers. The reactor temperature is between 840°F and 980°F at pressures between approximately 10 and 40 atmospheres. A lower reaction pressure results in high yields of liquid and hydrogen, whereas a higher pressure will produce the same yields at a faster rate, thus reducing the equipment size. Hydrogen is driven off the feed, forming a gas–liquid effluent. The effluent is sent from the reactor to a separator, where the separation of hydrogen and liquid occurs.

The hydrogen is compressed and fed back to mix with the feed at a hydrogen partial pressure of 5 to 35 atmospheres. The high pressure is necessary to avoid coke deposition on the catalyst. However, the catalyst can be regenerated either by removing the reactor for regeneration or by applying a continuously moving catalyst bed to the reactor. The liquid from the separator is guided to the fractionator to produce high-octane products.

Alkylation

Alkylation is the substitution of a substituent (such as propylene) for hydrogen (H) in the molecule of a hydrocarbon (such as isobutane). The most common alkylation reaction is that of isobutane, a saturated branched-chain hydrocarbon, with light olefins such as propylene or isobutylene.

$$
\underset{\text{Isobutane}}{\overset{\displaystyle \text{CH}_3}{\text{CH}_3 - \underset{\displaystyle \text{CH}_3}{\overset{|}{\text{CH}}}}} + \underset{\text{Propylene}}{\text{CH}_2 = \text{CH} - \text{CH}_3} \rightarrow \underset{\text{Isoheptane}}{\overset{\displaystyle \text{CH}_3}{\text{CH}_3 - \underset{\displaystyle \text{CH}_3}{\overset{|}{\text{C}}} - \text{CH}_2 - \text{CH}_2 - \text{CH}_3}}
$$

$$
\underset{\text{Isobutane}}{\overset{\displaystyle \text{CH}_3}{\text{CH}_3 - \underset{\displaystyle \text{CH}_3}{\overset{|}{\text{CH}}}}} + \underset{\text{Isobutylene}}{\text{CH}_2 = \overset{\displaystyle \text{CH}_3}{\overset{|}{\text{C}}} - \text{CH}_3} \rightarrow \underset{\text{Isooctane}}{\overset{\displaystyle \text{CH}_3}{\text{CH}_3 - \underset{\displaystyle \text{CH}_3}{\overset{|}{\text{C}}} - \text{CH}_2 - \overset{\displaystyle \text{CH}_3}{\overset{|}{\text{CH}}} - \text{CH}_3}}
$$

The products of this substitution are isoparaffins (isoheptane, isooctane), which have top antiknock qualities and are superior blending materials, especially for the low lead level gasoline. Alkylation improves the quality and increases the quantity of gasoline production.

The equipment required for the process includes a reactor–settler combined vessel, a fractionator, a depropanizer, a debutanizer, an HF (hydrofluoric acid) stripper, and an acid cooler, among others. Alkylation reactions are promoted either by heat alone or by catalysts. The thermal process requires high pressures and temperatures, which are added incentives to using the catalytic process.

The catalysts used in alkylation units are sulfuric acid (H_2SO_4) or hydrofluoric acid. When H_2SO_4 is used as a catalyst, the amount required and the waste generated are tremendously high. Also the reactor temperature must be controlled by using refrigeration equipment. In contrast, an HF catalyst can be regenerated economically, and the reactor temperature can be as high as 75°F to 115°F so that no refrigeration is needed. However, hydrofluoric acid is toxic, corrosive, and very volatile. In case of leaks or accidents, a deadly cloud may be formed; so special precautions must be taken in HF catalytic processes. The HF emissions usually are controlled by caustic scrubbers.

The feed for the HF process includes olefins from the catalytic cracking processes and makeup isobutane. The hydrocarbon feed mixes with hydrofluoric acid and enters the reactor at approximately 75°F. Hydrogen ion (H^+) initiates the reaction with the hydrocarbon mixture, forming liquid and vapor phases. The acid liquid is drained to the acid cooler and is recycled back to the reactor. The hydrocarbon vapor phase, which is a mixture of isobutane, propane, butane, HF acid vapor, and the alkylated product, is sent to the fractionator, where all the components are separated. Isobutane is recycled to mix with the feed; the remaining vapors are sent to the depropanizer, the debutanizer, and the HF stripper. The bottom yield is the product, the alkylate, an antiknock additive used in motor and aviation fuel.

The alkylated product also can be obtained through the sulfuric acid alkylation process. In this process, olefin and isobutane are fed to a reactor where sulfuric acid is injected. An impeller emulsifies these compounds, and the emulsion is sent to a settler, where the acid is separated from the hydrocarbon mixture and is returned to the feed. The hydrocarbon vapor is compressed, condensed, and cooled, and the cooled liquid is sent to a depropanizer to separate propane for recycling. The liquid hydrocarbons from the settler are pumped to caustic washes to neutralize the acid and then are guided into the isobutanizer. Isobutane vapor is removed, condensed, and recycled. Butane and the alkylate are produced.

Butane, a straight-chain hydrocarbon, can be rearranged to form a branched-chain hydrocarbon, isobutane, which is used as a feedstock for alkylation. This procedure, known as isomerization, is discussed below.

Isomerization
Isomerization, as used here, is the rearrangement of one hydrocarbon molecular structure to another geometrical structure with more desirable properties than those of the first structure. The procedure has been used to rearrange normal butane into isobutane, and normal hexane into 2,2-dimethylbutane; it converts

linear paraffins into branched isomers, which are used in motor fuel production. The octane numbers of the resultant isomers are improved by 8 to 10 points, depending upon the quality of the feedstock. Because unleaded gasoline is being marketed in an effort to reduce air pollution, the isomerization process has become particularly important. The main pieces of the equipment for the process are a heater, a reactor, an air cooler, a separator, and a fractionator.

The liquid feedstock can be butane, pentane, hexane, or their mixture. The feed combines with hydrogen (H_2) and is sent to the heater, where the feed temperature is raised to approximately 450°F to 525°F. The hot feed enters the reactor from the top and flows downward, passing through a fixed catalyst bed. The catalyst consists of platinum on a zeolite base and is regenerable; it has an acidic carrier support and a hydrogenation function. The reaction pressure is approximately between 13 and 30 atmospheres. The effluent from the reactor is air-cooled and is guided into the separator, where hydrogen is separated from the liquid phase and is returned to the feed. The liquid is transferred into the fractionator, where the bottom products are the highly branched isomers, and the top products are propane, ethane, hydrogen, and other coproducts.

Polymerization

Polymerization is the combination of small molecules of a compound to form larger molecules containing repeated units of the original molecular structure. The purpose of using this process in petroleum operations is to make the light olefin gases a gasoline liquid. Polymerization can be categorized into two processes: thermal polymerization and catalytic polymerization. Thermal polymerization is no longer popular because of its high temperature, high pressure, and low olefin conversion. Catalytic polymerization is used to produce gasoline components and diesel and jet fuels. The equipment for catalytic polymerization includes a reactor, a separator, and a stabilizer.

The catalyst, which is loaded on fixed beds inside the reactor, is phosphoric acid on pellets made of naturally occurring silica used as a carrier. The feedstock usually is propane or butane, a light olefin, which is heated to the required reactor-inlet temperature and enters the top of the reactor. The reaction temperature is approximately 375°F to 450°F, and the reaction pressure is between 40 and 80 atmospheres. A great amount of heat is liberated during the reaction; so the reactor temperature must be controlled by spraying a hydrocarbon liquid into the reactor. The reactor effluent is sent to the separator, where the gaseous components are separated from the liquid. The liquid is guided into the stabilizer, where the product, polygasoline, is obtained from the bottom end at the specified standard composition. The top end of the stabilizer yields propane or butane, which can be stored or recycled.

Because the octane number of polygasoline is not as good as that of the alkylate, polymerization is used only as a supplement when isobutane is unavailable or expensive.

SECONDARY OPERATIONS

When crude oil is converted into higher-quality petroleum products, impurities such as sulfuric acid, nitric acid, and undesirable compounds should be removed. Some processes remove the impurities from the feed (hydrotreating), and others remove them from the final products (gasoline treatment). These processes, as well as other support processes such as sulfur recovery and tail gas treatment, are secondary operations of petroleum refining.

Hydrotreating

Hydrotreating primarily is conducted with a catalyst to remove impurities in the feedstocks by hydrogenation. Through reaction with hydrogen, sulfur is converted to hydrogen sulfides, nitrogen to ammonia, oxygen to water vapor, and chlorine to hydrochloric acid. During the hydrotreating process, olefins are saturated, forming paraffins, for improved thermal and storage stability. The equipment required for the process includes a heater, a reactor, a separator, and a stripper.

The feed is mixed with hydrogen and is heated to 600°F to 800°F. The mixture enters from the top of the reactor, which is filled with a catalyst and has a pressure of between 40 and 60 atmospheres. For sulfur removal, a cobalt–molybdenum catalyst is used; for hydrogen saturation or nitrogen removal, a nickel–molybdenum is commonly used. The hydrogen consumption in cases of hydrogen saturation or nitrogen removal is more than two or four times, respectively, that of sulfur removal. The operating conditions are more severe than those of desulfurization. The reactor effluent is guided to the separator, where hydrogen is removed and recycled back to the feed. The liquid is sent to the stripper, where steam is injected to strip hydrosulfides and other undesirable gases. (Treatment methods for these gases are discussed below under "Sulfur Recovery" and "Tail Gas Treatment.") The liquid phase at the bottom of the stripper is the hydrotreated product.

Gasoline Treatment

Most gasoline products contain impurities, the amount and the type of the unwanted components depending on the quality of the feed. These impurities cause bad odors, corrosion, octane number reduction, and gum formation. The worst of these impurities are hydrogen sulfide mercaptans, thiophenols, and other organic acids. The purpose of gasoline treatment is mainly to remove the impurities from gasoline products to solve these problems. An additional advantage of the process is the conversion of harmful components to less harmful ones. The most common process uses a two-step principle, the reduction of mercaptans at the prewash and the oxidation of mercaptans in the final treatment. The equipment includes two sets of combined mixer–settler–coalescer devices.

In the prewash, the gasoline feed is combined with a weak caustic solution, containing approximately 5 percent by weight sodium hydroxide, and enters a mixer. The mixed liquid is sent to a settler, where caustic solution is separated from the gasoline and is recycled; then the gasoline liquid with entrained caustic solution passes through a coalescer, where the entrained solution is removed from the gasoline by a wool packing installed inside the coalescer. In the prewash stage, hydrogen sulfide, organic acids, and part of the mercaptans and thiophenols are removed.

The effluent from the first coalescer is combined with an aqueous caustic soda solution. The mixture is sent to the second mixer, where air is injected to promote oxidation of mercaptans and thiophenols to disulfides (S_2). The mixed product passes through the second settler and coalescer so that caustic solution is separated out, entrained solution is removed, and purified product is obtained.

Sulfur Recovery

The concentration of hydrogen sulfide (H_2S) in the exhaust streams of petroleum refining operations usually is very high; so the sulfur compounds are recovered primarily for economic reasons. The most common sulfur recovery method is the Claus process, which converts one part of H_2S in the stream to sulfur dioxide (SO_2) and then promotes reaction of the SO_2 product with the remaining H_2S in the presence of a catalyst, to yield sulfur. The equipment includes a burner, a waste-heat boiler, two converters, and two condensers.

The H_2S-rich stream is contacted with air in the burner and is partially burned to form SO_2:

$$2H_2S + 2O_2 \rightarrow SO_2 + S + 2H_2O$$

The combustion temperature is approximately 2200°F to 2600°F; the flue gas that is generated passes through the waste-heat boiler to generate steam. The flue gas contains SO_2 and H_2S, and its temperature is reduced drastically to approximately 450°F before it enters the first converter.

The catalyst is a natural aluminum mineral, bauxite, which is loaded on a fixed bed inside the converter to promote the following reaction between H_2S and SO_2:

$$2H_2S + SO_2 \rightarrow 3S + 2H_2O$$

The reaction is exothermic; so the effluent temperature is increased to about

560°F at the exit from the converter. The hot effluent is condensed to remove sulfur compounds, and the remaining effluent is sent through the second converter–condenser combination to recover additional sulfur.

This process can recover up to 95 percent of the sulfur. The remaining 5 percent is distributed as H_2S and SO_2, which are handled further as described below.

Tail Gas Treatment

Crude oil contains a high percent by weight of sulfur, which either evolves to H_2S gases or remains as a mercaptan (C_2H_5SH) or carbonyl sulfide (COS) in liquefied petroleum gases. These sulfur compounds are steam-stripped, forming H_2S gases. The carbon components are oxidized to CO_2. The two compounds of most concern in the tail gas are H_2S and CO_2; both form acidic aqueous solutions, and both compounds frequently are absorbed in regenerable solvents. The equipment includes an absorption column with trays, strippers, and regenerators. The solvents are different amines, such as diethanol amine, monoethanol amine, and diisopropanol amine. Diethanol amine is used mostly to remove H_2S from the tail gas, and monoethanol amine is used for CO_2 removal. The process, known as Girbotol absorption, is described briefly below.

The acidic gas is fed into the lower portion of the column and up through it. A basic solution, the amine, flows down through trays in the column, where it makes contact with and absorbs acidic H_2S and CO_2 from the countercurrent gases. Purified gas is vented from the top of the column into the atmosphere, while the acid-contaminated amine is sent from the bottom of the column to the stripper, where steam is injected into the amine stream. The stream temperature is raised to approximately 230°F, and the amine is stripped by the steam. The regenerated amine liquid is cooled to 100°F before being recycled back to the column. The gaseous stream is condensed in the regenerator to remove water vapor, and the remaining H_2S and CO_2 gases are sent to the Claus sulfur recovery unit. The overall sulfur recovery efficiency is approximately 99.9 percent. The last portion of the sulfur can be incinerated, flared, or treated further.

10

Nuclear Energy

This chapter briefly describes basic concepts of nuclear power, and then discusses nuclear power systems and their health effects. It concludes with a section on radioactive waste management and disposal.

NUCLEAR POWER FUNDAMENTALS

Nuclear power is associated with energy reserved in atoms, which are the smallest particles of elements that can exist alone or in combination. Atoms consist of protons, neutrons, and electrons. A proton is a positively charged mass located in the central part of the atom, known as the nucleus. A neutron, also located in the nucleus, has the same mass as the proton, but is not electrically charged. An electron, which has a negligible mass in comparison with a neutron or a proton, is negatively charged and revolves around the nucleus.

An atom generally possesses the same number of positively charged protons and negatively charged electrons; therefore, it is electrically neutral. The numbers of protons, neutrons, and electrons in atomic nuclei differ, depending upon the properties of the individual elements. Some atomic species have the same number of protons but a different number of neutrons, resulting in different atomic weights; these species are called isotopes of an element. Isotopes have the same physical and chemical properties, but different radioactive properties, because of their different number of neutrons.

For example, hydrogen, deuterium, and tritium are light element isotopes. Hydrogen (H) contains one proton in its nucleus. The nucleus of deuterium (D) contains one proton and one neutron; so its weight (atomic weight) is twice that of hydrogen. Tritium (T) has two neutrons and one proton in its nucleus; its weight is approximately three times that of hydrogen. When deuterium and tritium collide, two new entities are formed: helium-4 and a free neutron, which is accompanied by nuclear energy release.

Another source of nuclear energy is the heavy element extracted from uranium ore. The isotopes obtained from uranium ore include uranium-234 (U^{234}), uranium-235 (U^{235}), and uranium-238 (U^{238}). The amount of U^{234} is less than one-thousandth of one percent by weight of the uranium in the ore, and it is used mainly in nuclear research. U^{235} is the readily usable isotope of uranium that is integrated into nuclear fuel rods; however, the quantity of U^{235} in the ore is less than 0.1 percent. U^{238} is the most abundant isotope in uranium ore, constituting more than 99.9 percent of the uranium in the ore. U^{238} cannot be used directly for nuclear energy production, but it can capture a neutron in a nuclear reactor, forming plutonium-239, which can be used as nuclear reactor fuel.

The process that encompasses mining the uranium ore, producing reactor fuels, processing spent fuels, and disposing of radioactive waste is called the nuclear fuel cycle. In the mining process, uranium ore and dirt are excavated from the earth; and the dirt and the ore are separated at a milling plant. The milled products are sent to a process plant where U^{234}, U^{235}, and U^{238} are extracted. U^{235} is produced in an enriching plant, and uranium rods are fabricated in final form at a fuel fabrication plant. Then the fuel rods are ready for use in nuclear reactors. This stage is known as the front end of the nuclear power production process. At the back end, spent fuels are stored in temporary pools at the reactor site because some radioactive gases (for example, radon) are soluble in water. Some U^{235} can be recycled from the spent fuels to the enriching plant; some plutonium can be forwarded to the fuel fabrication plant; the remaining waste must be disposed of.

Between the two ends, nuclear power systems play an essential role. Major nuclear power systems, their impacts on human health, and radioactive waste management and disposal are described below.

NUCLEAR POWER SYSTEMS

The issue of nuclear power is relatively complicated, but its technology is fairly straightforward. A nuclear reaction produces either fission or fusion energy. The reaction processes, reactor types, and power systems for fission and fusion energies are quite different, as shown below.

Fission Power Plants

Nuclear power is generated by a radioactive substance undergoing fission in a reactor, which activates a power system. The following paragraphs give an overview of nuclear fission.

The Nuclear Fission Process

The nuclear fission process occurs when an atom of a radioactive element, such as uranium or plutonium, collides with a neutron and is split into two small parts accompanied by two or three neutrons. These neutrons can trigger further fission of nearby elements, causing a chain reaction. The neutron release is accompanied by the liberation of heat, and this thermal energy can be converted into electrical energy. Some neutrons are absorbed to trigger the fission process, whereas others escape or are captured by heavy elements without causing fission. In the latter case, other elements are produced in the reactor.

When the reactor product is fissionable and its amount exceeds that consumed in the fission process, the reactor is called a breeder. When the fissionable reactor product is less than the amount of consumed fissionable material, the reactor is known as a thermal reactor.

Nuclear Fission Reactors

A nuclear reactor is a device that can trigger a fission reaction and can be controlled for certain specific purposes. It consists of several major elements, from which different nuclear power systems are designed. The elements of a reactor include fuel rods, a moderator, control rods, a reflector, and a reaction vessel.

In the core of the reactor, fuel rods containing fuel pellets are installed. Each pellet contains approximately 3 percent fissionable materials, such as uranium-235, uranium-233, plutonium-239, or their combination. A moderator is a material used to decrease the speed of neutrons released by the fission process. The common moderating materials are ordinary (light) water (H_2O), heavy water (D_2O), and graphite. Control rods contain boron or hafnium, which can absorb neutrons. By inserting or withdrawing control rods into the core, the amount of neutrons and thus the fission rate can be controlled. The reflector is a metallic mirror surrounding the core used to reflect some of the neutrons, which otherwise would escape, back to the core. The reflector can be made of natural or depleted uranium.

The reaction vessel, which is sealed, contains the core, control rods, and coolants; the nuclear reaction is carried out inside it. The coolants absorb fission reaction heat from the core and transmit thermal energy from the reactors to various devices for power generation. The most common coolants are regular water, heavy water, liquid sodium, carbon dioxide, and helium. Reactors are designed for use in different nuclear power systems, described below.

Nuclear Fission Power Systems

Nuclear power systems are classified as direct and indirect cycles. They include thermal reactors, CANDU, and breeder systems.

Thermal Reactor System

A thermal reactor system performs fission reaction triggered by low-energy neutrons. When a neutron starts the fission reaction, regular water serves as the moderator and the coolant. The thermal energy released from the fission reaction is transmitted to the water.

In a direct cycle, the water in the reaction vessel is boiled and converted to steam, which is guided to drive a steam turbine connected to a power generator. The spent steam from the turbine is condensed, and the condensate (water) is pumped back to the vessel. The vessel and the water pump usually are confined in a concrete structure.

In an indirect cycle in a thermal reactor system, the vessel (primary) water absorbs the reaction heat, and the hot primary water is piped into a steam generator. After transferring heat energy to the secondary water in the steam generator, the primary water is pumped back to the vessel. The secondary water is boiled to form steam, which drives a steam turbine coupled with a power generator to produce electricity. The spent secondary water is cooled and pumped back to the steam generator to complete the cycle. Because the coolant, which passes through the reactor, does not flow through the turbine, this is considered an indirect cycle. The reactor, the steam generator, and the primary water pump are installed inside the concrete containment structure.

CANDU System

The CANDU system, or the Canada-Deuterium-Uranium reactor system, uses an indirect cycle with deuterium heavy water as the primary coolant and moderator, and uranium as the fuel. The heated primary coolant passes through the pressure tubes in the steam generator, and is pumped back to the vessel. The secondary coolant outside the tubes of the steam generator is regular water, which is converted to steam. Then the above-described steps are followed to generate electric power and to complete the cycle.

Breeder Reactor System

Plutonium or uranium can be used as the initial fuel in a breeder reactor. Subsequently, uranium-238 would be converted to fissionable plutonium during the breeder reaction. The primary coolant is liquid sodium, which becomes radioactive while it passes through the core. The radioactive liquid sodium transfers heat to nonradioactive liquid sodium at an intermediate heat exchanger, and the nonradioactive liquid sodium passes its thermal energy to a steam generator. Sodium and water pumps complete each individual cycle, and steam turbines drive power generators to produce electricity. This is an indirect cycle.

Fusion Power Plants

Fusion is the opposite of fission. In a fusion process, instead of splitting one heavy radioactive molecule into two smaller molecules, two light molecules are combined, forming a heavier molecule. Details are given below.

Nuclear Fusion

In nuclear fusion lighter elements with low molecular weights are combined to form heavier elements. In the process, neutrons are freed, and enormous amounts of energy are released. Deuterium and tritium are the major light elements used for fusion. The most common fusion reaction with these elements is the deuterium–tritium (D–T) reaction, whose products are helium-4 and a neutron. The energy released is approximately 14 MeV.

Another possible reaction is the fusion of two deuterium elements (D–D reaction), to form a tritium and a proton or a helium-3 and a neutron. The helium-3 can react further with another deuterium element, producing a helium-4 and a proton. The energy released from the latter (D–He 3) reaction is about the same order of magnitude as the D–T reaction. The energy yield from the D–D reaction, however, is only 20 percent that of the D–T reaction. These processes can be conducted in a fusion reactor.

Fusion Reactor

The basic components of a fusion reactor are a vacuum channel, a blanket, a thermal shield layer, a magnetic coil layer, and a biological shield layer. These components are assembled to form a horizontal, concentric, cylindric section with the vacuum channel at the center extended outward in the sequence blanket, thermal shield layer, magnetic coil, and biological shield layer.

The vacuum chamber is filled with deuterium and tritium elements (D–T fuels). The blanket, which consists mainly of the elements lithium-6 and lithium-7, will absorb neutrons and the released energy. The thermal shield layer is an insulator used to keep the reaction heat from being transferred from the blanket to the magnetic coils. The biological shield layer minimizes radioactive damage to the environment.

To initiate the fusion reaction, the D–T fuels are heated, either by injecting deuterium atoms or by inducing a high radio frequency. The injection of deuterium particles into the D–T fuels increases their kinetic energy; the high radio frequency has a microwave heating effect, increasing the orbiting frequency of the electrons. When the D–T fuels reach the plasma stage, the stream consists of electrons and ions. By charging the magnetic coils with an electric current, the plasma is concentrated at the central part of the channel and flows along the magnetic fields, causing collisions of D–T particles to form helium-4 and to release neutrons plus heat energy. The neutron energy thus released is absorbed by the blanket, to be transferred from the fusion reactor to devices associated with electric power generation. The neutrons also react with lithium in the blanket to breed tritium, which then joins in the D–T reaction.

There are two types of fusion reactors: the closed type and the open type. In the closed type the magnetic fields form a circular loop so that the energized plasma can flow along the loop continuously. In the open type both ends of the magnetic fields are interrupted by increasing the current intensity at the coils. Stable containment of plasma is possible with the open type of fusion reactor.

Fusion Power Systems

A fusion power system includes a reactor, a deuterium injector, a tritium separator, a heat exchanger, a steam generator, a steam turbine, a power generator, a condenser, and a cooling tower. The deuterium injector also can be used to inject D–T fuel into the reaction channel. The exhaust from the reactor contains various components, including tritiums, heliums, and electrons. These compounds are sent through the tritium separator, where only tritiums are allowed to pass through to the deuterium injector. The heat energy of the reaction is extracted by the primary coolant, passing through coils inserted in the blanket to the heat exchanger. The secondary coolant absorbs the heat energy in the heat exchanger, and transfers it to the steam generator to produce steam. The steam is used to drive a turbine to generate electricity. Spent steam is condensed, and the waste heat is treated in a cooling tower (Chapter 7).

Comparison between Fission and Fusion Energies

Several distinctions need to be made between nuclear fission and fusion energies. The primary difference between fission and fusion is that fission energy is released from the splitting of one heavy atom into two lighter ones, whereas fusion is the opposite, a combination of two lighter atoms to form a heavy one with the release of energy. The energy released from a unit fission reaction is greater than that obtained from fusion because two or three neutrons are freed from the fission reaction, but only one or no neutrons are released from the fusion reaction—although the release of a proton may free more energy than that obtained from a neutron. A fission reaction produces more radioactivity than a fusion reaction; for a D–D reaction especially, the production of radioactivity can be reduced to a minimum.

All fission reaction requires a radioactive substance, such as uranium or plutonium. All these fissionable substances are limited resources. The major fusionable substances are deuterium (which can be obtained from ocean water), tritium (which can be produced by the reaction of lithium and a neutron), and lithium (which can be extracted from soil). Thus the resources for fusion fuel are practically unlimited.

NUCLEAR POWER HEALTH EFFECTS

Uranium and other radioactive materials release radiation, which can have extremely adverse effects on personnel working with nuclear energy and radioactive materials, as well as on the general public. The characteristics of nuclear radiation are summarized below, and the biological effects of radiation and nuclear power health risks are briefly discussed.

Nuclear Radiation

A uranium fuel cycle includes ore mining and milling, fuel fabrication and transportation, reactor operation, and radioactive waste management and disposal. At each stage of the cycle, different types of radiation are emitted into the environment, the principal ones being neutrons, alpha particles, beta particles, and gamma radiation.

In ore mining and milling, the major radiation types are neutrons and alpha particles (positively charged nuclear material). The rate of energy transfer is very high when neutrons or alpha particles are intercepted (i.e., there is a high rate of linear energy transfer). Alpha particles present a high health risk. When an alpha particle hits the human body, it is stopped by the skin; it does not penetrate far into it. However, if the particle is inhaled, even as little as ten-millionths of a gram, and comes in contact with the lungs or other vital organs, it poses a cancer risk.

If a neutron is intercepted by human tissue, it will emit intensive gamma radiation to the surrounding tissue. Beta particles (positively charged electron or positron radiation) penetrate about one inch into tissue, and gamma radiation (high-energy photon radiation) requires several millimeters of lead to prevent its penetration. When radioactive particles penetrate tissue, a series of collisions occurs between the particles and the tissue atoms. Some electrons of the tissue atoms may be lost from their original structures, causing tissue ionization, molecular damage, and breakage of chemical bonds. The degree of ionization reflects the intensity of biological damage due to radiation.

The intensity of radiation is measured in rad units, where one rad is defined as 100 ergs per gram of tissue. The effect of radiation, however, is expressed in rem (Roentgen equivalent man) units. One rem is defined as a dose of any type of radiation causing a biological effect that is equivalent to that of one roentgen of X-ray or gamma-ray radiation. The intensity of one rad of radiation of beta particles or gamma rays causes approximately one rem of biological effect. One rad of neutrons or alpha particles, however, causes ten times or more the damage of one rem.

The average natural radiation in the United States has been reported to be 130 millirems per year, with manufactured radiation 80 millirems per year. Ninety percent of manufactured radiation comes from diagnostic procedures. The federal standards for radiation are that it must not exceed 170 millirems per year for average population exposure, and must be below 500 millirems for an individual. Present radiation-related activities will affect future radiation exposure, having adverse biological and environmental effects. This concept is known as dose commitment and is used in calculating the reference duration for the assessment of radiation-related health risks, discussed below.

Biological Effects of Radiation

The effects of radiation on human biology depend on many factors. Generally, the amount of radiation received by an individual woman will contribute significantly to birth defects in her children. Also, it has been reported that a whole-body dose of radiation from one to approximately 20 rems carries little risk of radiation-induced cancer; but with a whole-body dose of 20 to 200 rems, the radiation may have significant effects. A small region of DNA (deoxyribonucleic acid) molecules, which carry genetic information, can be altered; the materials in chromosomes can be rearranged; and the number of chromosomes can be changed. The consequences can be birth defects, cardiovascular disturbances, asthma, or diabetes.

The dose of radiation that will cause a mutation has been determined indirectly. The method defines a doubling dose as the amount of radiation that will double the incidence of a naturally occurring mutation. The naturally occurring human mutation rate was estimated to be $(0.5 \text{ to } 5) \times 10^{-6}$ per gene per generation. The average mutation per dose of radiation for mice was determined and reported to be 0.25×10^{-7} per gene per generation per rem. The human doubling dose then was defined as the ratio of these two values:

$$\text{Human doubling dose} = \frac{\text{Human natural - cause mutation rate}}{\text{(Mice) average mutation per dose}}$$

$$= \frac{(0.5 \text{ to } 5) \times 10^{-6}/\text{gene/generation}}{0.25 \times 10^{-7}/\text{gene/generation/rem}}$$

$$= 20 \text{ to } 200 \text{ rems}$$

This is the limit used to estimate the mutation effects of public exposure to a given radiation.

Nuclear Power Health Risk

The reference duration for this radiation-related risk analysis was chosen (by the Nuclear Regulatory Commission) to be a reactor year. One reactor year is equivalent to the radiation exposure received by individuals over a 50-year period due to power generation by a reactor for one year.

The risks of the nuclear fuel cycle were broken down as follows: Mining and construction worker accidental deaths per reactor year are estimated as 200 to 500 deaths per million of population, or 200 to $500/10^6$. The deaths due to occupational exposures during milling processes and reactor operation are estimated as 200 to $300/10^6$ deaths per reactor year. With an additional 200 deaths per reactor year due to routine exposure to radon radiation, the total health risk for nuclear energy related workers and the general public is 600 to $1000/10^6$ deaths per reactor year. Reprocessing and plutonium recycling may reduce the radiation from mining, milling, and tailings; however, emissions from transportation and fuel fabrication release high energy transfer radiation. In case of an accident, the casualties will depend on the affected population, its accident prevention readiness, and many other factors; this risk can be as great as ten or more times the total health risk for a population.

The majority of nuclear power health effects are associated with radioactive waste, the management and disposal of which are summarized in the following section.

RADIOACTIVE WASTE MANAGEMENT AND DISPOSAL

Radioactive Waste Categories

Radioactive waste comes from the following sources: hospital operation, research institutes, nuclear power plants, weapon systems, uranium mining, and milling. The wastes can be classified as high-level, intermediate-level, or low-level wastes, depending upon their activity levels. The activity level for radioactive waste is indicated by a unit of radioactivity called the curie (Ci), shown below:

$$1 \text{ curie} = 3.7 \times 10^{10} \text{ nuclear transformations/sec}$$

High-Level Radioactive Waste
High-level wastes are those with activity levels exceeding 10 curies per gram of radioactive materials. They include by-products from reprocessing reactor fuels and nuclear weapon systems. This class of waste generates very intense heat, and heavy shielding is required for handling the waste.

Intermediate-Level Radioactive Waste
Intermediate-level wastes are those with activity levels falling between 10 and 10^{-8} curies per gram of transuranic waste. The transuranic wastes include

elements with an atomic number higher than that of uranium (238), such as plutonium-239, americium-241, or curium-242. The heat generated by this class of waste is lower than that of high-level waste; however, proper shielding is required for handling the waste.

Low-Level Radioactive Waste
Low-level wastes are those with an activity level of less than 10^{-8} curie per gram of radioactive material. This category includes radiopharmaceuticals, chemicals, contaminated laboratory equipment, clothing, and tools from research laboratories, hospitals, and industries. The waste, with an approximately 80 to 90 percent organic content, generally is untreated. It consists of wood, cotton cloth, rubber, plastics, metals, ceramics, glass, and dead animals.

All categories of radioactive waste need to be handled properly, but final decisions are hard to reach because of serious problems affecting radioactive waste management policies.

Radioactive Waste Management

The two major concerns in radioactive waste management are high- and intermediate-level wastes and the tailings from mining and milling processes.

High- and Intermediate-Level Waste Management
Spent fuels and the reprocessing and recycling of uranium and plutonium generate the most high- and intermediate-level radioactive waste. When spent fuels are removed from a nuclear reactor, the intensity of the radioactive emissions initially is so great that the radioactivity will be reduced by a factor of 1000 in the first 10 years. The fission products include plutonium, uranium, strontium-90, technetium-99, iodine-129, and many other isotopes. Strontium-90 shows up in milk obtained from the surrounding areas, technetium-99 in stomach and intestinal systems, and iodine-129 in the thyroid.

Only a small portion of spent fuels is reprocessed, with the plutonium and uranium recycled. The majority of spent fuels are temporarily stored in water-cooled facilities near reactor sites, awaiting a final decision on whether they should be either permanently disposed of or reprocessed for reuse. The reprocessing of spent fuels produces liquid and gaseous emissions, and generates solid wastes. The liquid emission is mainly an acidic stream that contains plutonium and uranium. The gaseous emissions are krypton-85, iodine-131, and carbon-14; krypton and iodine emit beta particles and gamma rays, and carbon-14 releases beta particles. The solid waste contains plutonium; when the waste is buried in shallow land, it has been reported that the plutonium will migrate several hundred meters in several years. All these wastes should be either treated, secured, and retrievably stored or permanently disposed of.

The elements in the gaseous emissions have various properties. For example, krypton is slightly soluble in water and can combine with fluorine at the temperature of liquid nitrogen (–346°F); iodine is soluble in alkaline solution; and carbon is emitted as carbon dioxide, which can be controlled by using amine scrubbers. The acidic liquid waste can be solidified to remove most of the water, forming a salt cake, or can be sprayed into a fluidized bed, forming a calcine. The latter procedure will reduce the waste's volume greatly; moreover, the residue can be immobilized in glass or ceramics. From a health standpoint, salt cake removal is risky and costly. It is uncertain whether the seals used to immobilize a calcine are permanent, and whether immobilized land can be kept vacant for a long period of time.

Low-Level Waste Management
The residue from the mining of uranium rocks or the milling of uranium ore (tailings) contains radioactive substances. The concentrations of the radioactive substances are low, but the radiation emitted from the tailings are alpha particles, which pose a high cancer risk. Because of the enormous quantity of the tailings, these residues are piled in open-surface areas and are scattered by erosion and wind, creating vexing problems. Proper control measures are either nonexistent or minimal, but the disposal of low-level wastes is as important as that of high-level wastes.

Radioactive Waste Disposal

The methods used for radioactive waste disposal differ, depending upon the levels of radioactivity of the waste. For all levels of waste, methods of waste retrieval and permanent disposal methods have been proposed; but for industrial low-level wastes, permanent disposal has been the preferred approach.

High- and Intermediate-Level Waste Disposal
The proposed means of retrievable disposal for high- and intermediate-level wastes is to isolate the radioactive waste by burying it in a container in a repository until it becomes necessary to retrieve the waste to reprocess and recycle nuclear fuels. A practical site should be chosen carefully; it should not be in any area of volcanic activity, flooding, earthquakes, or mining, nor should it be in a war zone or an area easily accessible to terrorists or saboteurs.

In containers (steel casks) that have been proposed by Bechtel National, Inc., spent fuels are assembled in the central part of the cask, where they are separated by support baskets. The outside of the cask consists of a steel shielding material, an impact limiter, and a neutron shielding material in the outermost layer. The cask is transported to an underground facility; and after its closure is checked, the cask is raised and rotated and placed vertically in the isolated repository. The major concerns of this method are that any improper design may cause a container seal failure; undiscovered holes may induce a future intrusion into the container; thermal and radiation forces may increase internal pressures, moving the container and causing seal failure. Therefore, a cask must be able to withstand: severe accidental impacts, a free-fall from a specified height, being struck at the weakest point on a specified steel nipple, fire for a specified period of time, and water immersion for a specified duration.

Once these criteria are met, the containers can be retrieved when the waste needs to be reprocessed, or when the containers need to be inspected, reinforced, and eventually removed for permanent disposal. For high- and intermediate-level wastes, the following processes have been suggested and discussed by numerous scientists: geological formation on land disposal, an in situ underground melt process, ocean dumping and subseabed disposal, ice sheet disposal, and extraterrestrial disposal. These methods have not yet been put into practice.

Geological Formation on Land Disposal

In this proposed process, radioactive waste is sent to a receiving facility connected by vertical shafts to a salt mine that is approximately 2000 feet underneath the ground, far below the groundwater. The waste is placed in canisters, which are transferred along the shafts to carts that move the waste to its storage place. Generally stainless steel containers serve as barriers to leakage. However, when the barriers degrade, risks arise depending upon the physical situation: any wastewater present will carry radioactive contaminants and transfer them to the environment; leachable waste will contaminate and migrate to surrounding layers; permeable surrounding rock will contribute to further transmission of radioactive substances.

Salt beds have been suggested for use as repositories because of their location and salt characteristics. The beds are located in a layer free of groundwater, as they are isolated from groundwater by an impermeable shale. They are located in areas of low seismic activity and are available worldwide. Because of its high thermal conductivity and structural strength, salt can remove heat from waste containers and withstand radiation. If there are any fractures of salt domes due to pressure or temperature, salt can heel the cracks by flowing plastically into the gaps, deforming, and recrystallizing. A dangerous disadvantage is that salt is corrosive to containers. It should be guaranteed that human activities will not interfere with the salt dome repository.

In Situ Underground Melt Process

The proposed in situ underground melt process pumps high-level liquid into an underground "well" where the waste originated. The well can be a hollow space, 3000 feet deep, formed by an underground nuclear detonation. While the high-level liquid waste is being pumped into the well, steam vapor is released from the well upward, and it should be treated before being emitted into the atmosphere. Rubble is added to the waste, and after a certain period of time, the entrance to the well is sealed. The rubble begins to melt, as do the surrounding rocks. When the melt reaches its maximum radius, the rock begins to be refrozen. The unknowns in this process are radioactive vapor control, geographical changes, groundwater effects, the techniques, and the consequences of the process.

Ocean Dumping and Subseabed Disposal

Waste is to be put into canisters that can be either dumped into the ocean or put inside projectiles and propelled through the water to be placed firmly within the subseabed. The advantages of this method are that it gets the waste out of sight, disposes of it in a stable place, cools the generated heat with sufficient water, and has perfect seals. There are questions about the uncertainty of geographical effects, as well as effects on the ocean's resources and their future uses. Also seawater is corrosive; and when the canisters are broken, the wastes will mix with the upper water layer and enter the food chain.

Ice Sheet Disposal

This method would dispose of high-level waste-filled containers underneath the South Pole (Antarctic) ice sheet. It has been estimated that the ice sheets undergo relatively rapid movement every 10,000 years, causing large-mass water flow and violent waves. If the ice disposal method is used, heat released from the waste may cause additional violent movement, creating a hazard for human activities.

Extraterrestrial Disposal

This is a proposed method for disposing of solidified high-level waste in extraterrestrial orbits. The waste containers would be transported by using a space shuttle and releasing the contained waste in a predetermined orbit between Earth and Venus. Its advantages would be moving the waste away from Earth and having stable orbits for at least several million years. However, the number of shuttle trips is limited, and the risk of malfunction of a shuttle before going into orbit is a major concern. The cost has been estimated as $1000 per pound of payload; so the costs alone could make the method unfeasible.

All the above methods have been proposed specifically for high- and intermediate-level wastes. Low-level waste mainly is disposed of in shallow land trenches or aboveground tombs. The site selection, design criteria, and configuration are particularly important, and special precautions are necessary.

Low-Level Waste Disposal

The site for low-level waste disposal should be a location where land is stable and relatively unsusceptible to earthquakes and landslides. The population density around the area should be sparse, and the site should be isolated. Major design criteria for a trench or a tumulus module are as follows:

- The module should be above groundwater.
- The land use should be optimized.
- The module should withstand both dry and wet conditions.
- It should be possible to pump out the water accumulated inside the module.
- The module should last as long as the life of the radionuclide (estimated to be 400 years), or the module should be retrievable every 20 years.

Shallow Land Trench Configuration

A shallow land trench usually is set at a depressed geographical location. The floor of the trench is slightly sloped so that the water inside the trench can be accumulated in a sump and pumped out for further treatment. A rubber or plastic lining is spread on the floor and the outside wall of the trench module, and dirt boxes are placed along the wall next to the lining. Radioactive waste is deposited in the space between the dirt boxes. An impervious liner covers the waste at the top of the trench module, and a cap is put on top of it. A drainage lining is added to the cap before the compacted dirt is backfilled. A gas vent may be required for the module.

Above-Ground Tumulus Configuration

When a depressed geographical location is not available, an aboveground tomb can be used. On the ground, a gravel base and gravel backfill are set inside the module; and a reinforced concrete floor is on top of the gravel floor. Sumps also are provided on the concrete floor. The outer walls are stacked with concrete drums; and radioactive waste is deposited between partitions that are also concrete drums. After the gaps are filled with gravel, clay is layered on top of the waste–gravel tomb structure. Several layers of material are spread on the clay layer, in a sequence of sand, plastic liner, soil, gravel, and topsoil.

Both the trench and the tumulus modules are for low-level waste disposal only. If water in the modules cannot be removed, an additional liner(s) would be needed, and an alarm would have to be installed.

11

Major Alternative Fuels and Advanced Technology

In addition to the conventional fossil fuels and nuclear energy, several alternative fuels are under development and have partially proved to be feasible in large-scale applications. The major alternatives include liquefied petroleum gas (LPG), compressed natural gas (CNG), methanol, ethanol, and hydrogen. Also some advanced technologies, such as fuel cells and superconductors, are effective and efficient with respect to energy consumption. Most important, the amount of pollution caused by these fuels is limited if not negligible.

ALTERNATIVE FUELS

Gasoline and diesel fuel are environmentally unattractive fuels, and their use as a future energy supply, along with its economic implications, is unknown. Alternatives to these fuels have undergone intense and fruitful investigation. Some of these alternatives are discussed below.

Liquefied Petroleum Gas

Liquefied petroleum gas (LPG) is a generic name for propane, butane, or a mixture of both. There are two means of producing LPG: the petroleum LPG process and natural gas stream extraction. The petroleum process uses crude oil as a liquid feed, whereas natural gas stream extraction uses natural gas liquid as the source. (There has been a decline in the use of refinery crude; the extraction of LPG from the natural gas stream is increasing.)

In the petroleum LPG process, the liquid feed is guided to an LPG unit, which consists of a deethanizer column and a depropanizer column. The deethanizer is heated by a reboiler, a heating unit supplying additional heat to other equipment. The overhead vapors are cooled by a condenser and a reflux drum, where the ethane (gas) and the liquid are separated, following the deethanizer. The ethane gas is treated further and stored, and the liquid is pumped back to the column. The remaining liquid in the column is sent to the depropanizer column, where the propane is vaporized from the top of the column; again it is condensed and removed from the system. The remainder in the second column is butane.

Natural gas from a well consists of methane, natural gas liquid (NGL), carbon dioxide, nitrogen, nonhydrocarbons, and water. The NGL can be separated in fractionators into the following substances: ethane (C_2H_6), propane (C_3H_8), butane (C_4H_{10}), pentane (C_5H_{12}), and other heavier hydrocarbons. At atmospheric pressure and standard temperature, propane and butane are in the gaseous state. They can be liquefied by either compression or refrigeration, or both.

To liquefy propane and butane, the pressure is increased to approximately 8 and 2 atmospheres at standard temperature, respectively, and the volumes of the liquids are reduced, about 270 times for propane and 240 times for butane. The LPG also can be obtained by cooling the gases down below their boiling points at atmospheric pressure, to approximately $-45°C$ for propane and $-1°C$ for butane. The LPG then is put into either pressurized cylinders or small-bulk pressure vessels, to be transported to application sites. During the application, the pressure of the LPG is reduced so that it vaporizes, and the resultant gas is piped away to a desired place for use. The combustion of LPG is relatively clean; there is neither smoke nor residue. The LPG is noncorrosive, and adds little pollution to the atmosphere. For a larger amount of LPG usage, a pump is required to withdraw LPG from the bulk tank to a vaporizer before it reaches the point of application. When the pressure is released, LPG is vaporized; then the heat of vaporization is absorbed by the surroundings so that the temperature of the LPG drops below the boiling point, preventing LPG gas formation.

LPG is heavily used because of its low level of exhaust emission. It is used for cooking and heating in residential and commercial applications, for a high-quality fuel in industrial settings, and for an automobile fuel in urban transportation. A drawback is that pressurized cylinders and bulk vessels are expensive; also it is difficult to store large quantities of LPG in pressurized tanks because of technical limitations and because it is not cost-effective. Underground caverns may prove to be an option for storing large amounts of LPG; however, suitable rock structures are not always available for this particular purpose.

Compressed Natural Gas

Natural gas has been used in stationary and mobile industries for decades. It has been used to serve as a feedstock instead of some petroleum product, to replace fuel oil in utility power plants, and in industrial fuel applications. An important contribution of natural gas is that it has been compressed to as much as 3000 pounds per square inch (psi) and used as a vehicle fuel.

Compressed natural gas (CNG) is transferred in a high-pressure cylinder on board the vehicle. When a master valve is turned on, CNG travels through a high-pressure fuel regulator in the engine compartment. The natural gas then enters the carburetor at atmospheric pressure, and there is mixed with air at a proper ratio. The gas mixture flows into the engine combustion chamber, where it is ignited to deliver power to the vehicle.

CNG has a high octane value, approximately 130. The engine it powers has greater efficiency, less engine block, longer engine and spark plug lives, and longer periods between oil changes, in comparison with its gasoline engine counterpart. CNG vehicles also have excellent cold-start and hot-weather drive characteristics. Although CNG vehicles will lose approximately 10 percent of their power in the spark-ignited engine (because CNG expels oxygen in the combustion chamber), CNG diesel engines increase their power at a higher compression ratio (14 to 1). This may help to explain why CNG-conversions are concentrated on transit buses, fleet trucks, and heavy-duty diesel engines.

In the United States, there were more than 30,000 natural gas fueled vehicles, served by about 300 refueling stations in 1988. Worldwide it was reported that 500,000 vehicles had been converted to natural gas vehicles, serviced by 800 refueling stations. These figures may be expected to increase dramatically because of rapid urbanization, strong growth in motorized vehicles, air-quality deterioration, and imported oil–induced trade imbalances.

The following are specific benefits of using CNG as a vehicular fuel, besides the advantages involving national trade balances and energy security:

- It is relatively inexpensive.
- It is safer than other motor fuels.
- It requires relatively little vehicle maintenance.
- It decreases the likelihood of fuel theft.
- There is little chance of an interruption to its supply.

Natural gas is an abundant resource. At today's comsumption rates, it has been estimated that there is a 200-year supply worldwide. The possibility of any natural gas shortage in the foreseeable future is not a relevant issue at this time. The significance of using CNG as a vehicular fuel is in environmental considerations, including criteria pollutants, air toxics, global warming, and safety concerns.

Various data have been reported on the reduction of criteria pollutants emitted from CNG-converted vehicles. The percentages of specific emission reductions depend upon many factors, such as tuning, age, vehicular design, the condition of the gasoline emission controls, and the CNG conversion kit. Roughly speaking, a CNG-converted vehicle should show a reduction of at least 85 percent of the reactive hydrocarbon emissions and 80 percent of the carbon monoxide. The emission of nitrogen oxides is about the same as that of conventional gasoline or diesel-powered vehicles.

The air toxics emitted from conventional vehicles are aromatic hydrocarbons, polycyclic aromatic hydrocarbons, and formaldehyde. The aromatic hydrocarbons include benzene and toluene emitted through evaporation and from the exhaust. Polycyclic hydrocarbons are emitted as particles from diesel exhaust. Formaldehyde is formed in the combustion of petroleum products. All these compounds are known carcinogens, which need to be controlled before being emitted to the atmosphere. The emissions of aromatic hydrocarbons and polycyclic hydrocarbons are reduced substantially by using CNG; however, the emission of formaldehyde is about the same as that from gasoline vehicles.

Other by-products of the combustion process are NO_x, CO_2, CO, and unburned hydrocarbons, such as CH_4. NO_x is a precursor of ozone formation that, together with the cited compounds, constitutes greenhouse gas, which causes global warming problems. It was reported that the NO_x emissions from engines designed for natural gas were 70 percent less than those from gasoline-fueled heavy-duty engines in 1991. Also the use of CNG vehicles causes a reduction of 25 to 30 percent of the carbon dioxide emissions; and when CNG is adjusted in lean burning, carbon monoxide emissions can be eliminated from the exhaust gases.

One can conclude that CNG is less toxic, less groundwater-polluting, and less corrosive than gasoline. From a safety standpoint, natural gas is safer than gasoline for the following reasons: (1) natural gas is lighter than air, so that when it leaks, it is likely not to accumulate at one spot to cause danger; and (2) natural gas has a high ignition temperature (approximately 1000°F) and ignites only in a narrow range of the air/gas ratio.

The disadvantages of using CNG as a vehicular fuel are that refueling stations are not so popular as gasoline stations, and an extended time is required for CNG refill. These problems are expected to be solved, at least partially, by further technological developments.

The Gas Research Institute in the United States has initiated a series of new projects related to CNG-powered vehicles, including the following:

- A diesel transit bus conversion kit for dual fuel use (natural gas/diesel) or for CNG only.
- A CNG four-cycle engine for transit buses.
- A CNG-fueled engine for refuse trucks.
- A CNG-fueled engine for light-duty four-cycle engines.
- A CNG-storage system for 500 psig (instead of 3000 psig) with a high storage density.
- Technology to fill a vehicle's CNG tank to 3000 psig in 5 minutes or less.

With all these efforts, CNG can be expected to be a competitive vehicular fuel worldwide.

Methanol Fuel

Methanol (CH_3OH) is known as wood alcohol or wood spirits because it initially was produced by destructive distillation of wood. It also can be obtained by direct thermal oxidation of methane with oxygen, or can be synthesized from hydrogen and carbon monoxide at a temperature of 300 to 375°C and a pressure of 270 to 350 atmospheres. Carbon dioxide can be used in place of carbon monoxide to produce methanol.

The reaction formulae are the following:

$$CH_4 + \tfrac{1}{2}O_2 \rightarrow CH_3OH$$

$$CO + 2H_2 \rightarrow CH_3OH$$

$$CO_2 + 3H_2 \rightarrow CH_3OH + H_2O$$

Unless the methane is oxidized, the above mixtures are purified to remove sulfur and are compressed to 100 to 600 atmospheres. The compressed mixtures pass over and recirculate around a catalyst made of cobalt, manganese, titanium, and zinc, at a temperature between 250 and 400°C. The resulting vapors are condensed to form a liquid that is removed and purified, yielding methanol.

Methanol is highly toxic; its injection or inhalation causes blindness or death. The threshold limit value for exposure to methanol averaged over an 8-hour workday is given as 260 mg per cubic meter by the American Conference of Governmental Industrial Hygienists, Inc. (ACGIH). A single exposure of less than 15 minutes is recommended not to exceed 200 mg per cubic meter; the worst-case exposure should be less than 1 mg of methanol per kilogram of the exposed person's body weight.

Patrick Kinney and Robert Kavet found that toxic exposure to methanol will result, initially, in a depression of the central nervous system. This may damage the nervous system, causing Huntington's disease. Further, they reported the following: the interval between the time of the exposure and the first appearance of the symptoms is commonly 12 to 24 hours. The body's pH becomes extremely low, having a toxic effect on the visual system, which is lethal if untreated. Therefore, methanol should be stored and handled with care. Methanol has a high vapor pressure, 99 mm Hg at 20°C; it is flammable, forming explosive mixtures in closed containers. The installation of a vapor recovery system has been recommended for the methanol storage, fueling, and dispensing facility.

Methanol can be used for the production of resins and chemicals. Also it can be used as an anti-detonant fuel-injection fluid, as a fuel in modified engines for racing cars, and in mini-power plants. It also is a fuel additive for rocket, jet, and combustion engines. As an energy source, methanol has been applied in a neat form as well as a blend of alcohol and gasoline. The major exhaust gases are NO_x, CO, and HC from the unburned fuel. With the same stoichiometric fuel-to-air ratio, a methanol combusion product contains less NO_x than that from gasoline because the combustion temperature of methanol is lower than that of gasoline. CO and HC emissions can be decreased by burning methanol with excess air.

Generally, ozone and particulate matter are decreased by burning methanol; however, methanol leakage and formaldehyde (CH_2O) formation are high. Formaldehyde is a by-product of the incomplete combustion of methanol. Also it has been reported that in the oxidation of methanol, formic acid (CH_2O_2) is formed, which is immediately dissolved to formate (COO^{-2}) and hydrogen ion (H^+). The formate is toxic, and its accumulation will damage the visual organs. Sources of methanol vapor emissions include unburned material in the vehicle's exhaust, refueling evaporation, carburetor emissions following ignition turnoff, and heating of the fuel tank to overcome cold-start problems. At approximately $-10°C$, it is difficult to form a flammable air mixture with methanol.

The cold-start problem can be solved by any of the following methods:

- Addition of more volatile fuels to the methanol.
- Operation of an auxiliary starting system using other fuels.
- Electrical heating of the fuel tank.
- The use of a catalytic methanol reformer to produce hydrogen to assist the engine in starting.

It has been reported that addition of 10 to 20 percent water to methanol raises its octane value from 102 to 107, and the thermal efficiency of aqueous methanol can be expected to be higher than that of the neat methanol.

Methanol can be blended with gasoline for use as a fuel. The primary reason for this is to increase vaporization, to increase engine performance, and to substitute a certain portion of the gasoline. At a concentration of 10 percent by weight in gasoline blends, methanol evaporates at approximately twice its normal vaporization rate. When the temperature of the fuel exceeds the boiling point of methanol, the fuel vaporizes and forms air bubbles in the fuel system, stopping the flow of fuel to the engine and interrupting engine operation. Once the volatility balance is disturbed, vapor lock will occur.

Methanol–gasoline blends are miscible with water. Water contamination can cause phase separation of the blends into two layers: an upper hydrocarbon-rich layer and a lower water-rich layer. The upper layer's fuel blend has a lower alcohol concentration that can adversely affect engine operation, whereas the lower water-rich layer can cause corrosion of the fuel system. Therefore, long-term storage of methanol–gasoline blends is not recommended, because of absorption of moisture from the air. Additional defects include an increased rate of gum formation and a loosening of existing deposits of rust by methanol, which plugs the fuel system and decreases engine efficiency.

Ethanol Fuel

Ethanol (C_2H_5OH) or ethyl alcohol, is consumed in alcoholic beverages. It is a volatile, flammable, clear, and colorless liquid, with a pleasant odor and taste. It can be produced by fermentation of biomass, derived from ethylene, or synthesized.

The biomass used as a basic source of ethanol consists mainly of starchy crops and sugar crops. The starchy crops are ground, slurried with water, and anaerobically treated with dilute acid to liberate sugars, which then are fermented with yeasts. The stoichiometric reactions are as follows:

$$C_6H_{10}O_5 + H_2O \rightarrow C_6H_{12}O_6$$

$$C_6H_{12}O_6 \rightarrow 2C_2H_5OH + 2CO_2$$

The end products are ethanol and carbon dioxide.

Ethanol can be derived from ethylene synthetically through indirect hydration or direct hydration processes. The indirect hydration process requires the presence of dilute acids, whereas direct hydration eliminates the use of the acids.

Indirect Hydration Process
This is a three-step process: absorption, hydration, and separation.

Absorption Step
Ethylene is passed through highly concentrated (98%) sulfuric acid in absorbers; the high concentrations of sulfuric acid can improve the absorption effects. This is an exothermic reaction, and the absorbers require cooling to avoid corrosion. The product of the reaction, the absobate, is a mixture of monoethyl sulfate and diethyl sulfate. The formulae for the reactions are as follows:

$$(CH_2)_2 + H_2SO_4 \rightarrow CH_3CH_2OSO_3H$$
ethylene   monoethyl sulfate

$$2(CH_2)_2 + H_2SO_4 \rightarrow (CH_3CH_2O)_2SO_2$$
diethyl sulfate

Hydration Step
The absobate, the mixture of monoethyl and diethyl sulfates, is hydrolyzed with enough water in the hydrolyzer to yield ethanol and a dilute aqueous sulfuric

acid solution of approximately 50 percent concentration. The chemical reactions are as follows:

$$CH_3CH_2OSO_3H + H_2O \rightarrow CH_3CH_2OH + H_2SO_4$$

monoethyl sulfate   ethanol

$$(CH_3CH_2O)_2SO_2 + H_2O \rightarrow 2CH_3CH_2OH + H_2SO_4$$

diethyl sulfate   ethanol

Diethyl sulfate will react with ethanol to form monoethyl sulfate and diethyl ether; therefore, the diethyl sulfate needs to be removed from the final product (ethanol). Diethyl sulfate is harder to hydrolyze to ethanol than monoethyl sulfate; so diethyl sulfate should be removed prior to hydration. The formation of diethyl ether can be seen from the following reaction:

$$(CH_3CH_2O)_2SO_2 + CH_3CH_2OH \rightarrow CH_3CH_2OSO_3H + (CH_3CH_2)_2O$$

diethyl sulfate   ethanol   monoethyl sulfate   diethyl ether

Separation Step
The mixture of the ethyl sulfates and the sulfuric solution is guided to stripping columns for separation of the gaseous mixture, on top, from the liquid sulfuric solution at the bottom of the columns. The gaseous mixture thus collected is the desired crude ethanol.

Direct Hydration Process
The direct hydration process guides the vapor-phase ethylene over a solid or a liquid catalyst. Here the catalyst is a support impregnated with phosphoric acid, which is used to assist chemical reactions at low temperatures. By supplying phosphoric acid to the catalyst during the reaction, the lifespan of the catalyst can be extended. The reaction formula for the direct hydration process of ethylene is as follows:

$$(CH_2)_2 + H_2O \xrightarrow{\text{Catalyst}} C_2H_5OH$$

ethylene   ethanol

The reaction can be performed in a reactor (Union Carbide method), where the product is cooled and separated into liquid and gaseous phases. The liquid phase can be distillated to yield the product, ethanol. The gaseous phase contains ethylene, which is recycled back to the reactor as a feedstock.

The quantity of ethanol produced by the synthetic method is far greater than that from fermentation. Another synthetic method includes carbon monoxide and hydrogen as reactants, using the formula below:

$$2CO + 4H_2 \xrightarrow{\text{Catalyst}} C_2H_5OH + H_2O$$

ethanol

The catalyst for this reaction is powdered iron instead of the phosphoric acid or sulfuric acid used above for the hydration methods. The conditions for the reaction are approximately 150°C and 14 atmospheres. The advantage of using this method is that the reactants, carbon monoxide and hydrogen, are available from municipal waste treatment plants, or biomass conversion facilities, which is an environmental benefit.

Besides the danger of its flammability, ethanol vapor is explosive; also it causes eye irritation and is fatal with 500 g ingestion. Its threshold limit value has been set at 1000 ppm by the American Conference of Governmental Industrial Hygienists.

In addition to its applications as a solvent, a germicide, a beverage, an antifreeze, a depressant, and an intermediate for making other organic chemicals, ethanol can be used as a motor fuel, either in a neat form or as an ethanol–gasoline blend. Neat ethanol has a high latent heat of vaporization, 839 joules/g, approximately double that of unleaded regular gasoline (349 joules/g). When it is used as a motor fuel, the fuel evaporates partially during the suction stroke and continues to evaporate during the compression stroke, resulting in a high power output. The stoichiometric combustion of neat ethanol is shown below:

$$C_2H_5OH + 3O_2 \rightarrow 2CO_2 + 3H_2O$$

Neat ethanol–fueled cars were introduced in Brazil well over a decade ago. It was reported that with an approximately 15 percent by weight water content, ethanol increases its research octane number by approximately 10 percent. A cold-start problem arises at temperatures below 15°C for neat ethanol–fueled cars. The problem has been solved, however, by using a gasoline-assisted engine starter.

In a blend of gasoline with 10 percent by weight ethanol, the evaporation rate of the ethanol increases approximately 50 percent, and the vapor pressure of the blend is generally higher than that of gasoline alone. Similar gas-lock problems may occur in the fueling system for methanol–gasoline blends. The ethanol-blend fuel has other characteristics similar to those of the methanol blends, such as corrosion, orifice plugging, and phase-separation problems; but the ethanol blend is less sensitive to phase separation than are the methanol blends.

Generally, for the same distance a vehicle consumes approximately 20 percent more ethanol or 50 percent more methanol than gasoline. In the future, the energy cost of ethanol could be less than that of methanol; and because of engine performance, ethanol can be a competitive fuel source compared with gasoline and methanol.

Hydrogen Fuel

The hydrogen atom is comprised of a single proton as the nucleus and a single electron orbiting about the nucleus. Hydrogen gas, H_2, is the lightest gas known, and it is a very stable molecule because of its high binding energy (104 kcal/mol). Under standard conditions it is unreactive, but it will undergo many reactions with the aid of a catalyst or at elevated temperatures. In the presence of a platinum or nickel catalyst, hydrogen converts aldehydes to alcohols.

At high temperatures, hydrogen will react with many elements. Also, when hydrogen expands rapidly from a high to a low pressure, it heats up and burns rapidly, with a flame temperature exceeding 2000°C. Neither carbon monoxide nor other fumes are produced; the exhaust is mainly water vapor. A mixture of hydrogen and a flammable gas is explosive when a spark or a flame is present, but the volume of the explosive mixture is relatively small.

Hydrogen can be produced by various methods, such as steam-reforming of hydrocarbons, partial oxidation of hydrocarbons, water electrolysis, coal gasification, and various laboratory-scale methods. Industrial hydrogen is produced by steam reforming and partial oxidation methods.

The method of steam-reforming of hydrocarbons mixes a vaporized hydrocarbon with steam and passes the mixture over a nickel catalyst at 1600°F, producing hydrogen as indicated in the following reaction for methane reforming:

$$CH_4 + 2H_2O \xrightarrow{\text{Ni Catalyst}} CO_2 + 4H_2$$

In the partial-oxidation method, atomized hydrocarbons are contacted with a steam and oxygen mixture. The hydrocarbons react partially with oxygen, forming water vapor and carbon dioxide; then the remaining hydrocarbons react with the water vapor and steam to produce hydrogen and carbon monoxide, which reacts further with water vapor to yield hydrogen and carbon dioxide. The feedstocks can be any hydrocarbon, ranging from natural gas to crude oil or asphalts. The overall reaction is as follows:

Hydrocarbon + Water + Oxygen → Hydrogen + Carbon dioxide

In the electrolysis method, an electrical current is conducted through a water solution, splitting the water into hydrogen and oxygen ions in an electrolysis cell. The cell consists of a tank, a cathode, an anode, an electrolyte solution, and a porous diaphragm. When water dissociates into a hydrogen ion, H^+, and a hydroxyl ion OH^-, the hydrogen ion moves to the cathode, and the hydroxyl ion flows to the anode. The ions then are discharged from the electrodes, releasing hydrogen and oxygen gases. The porous diaphragm separates the two electrodes and prevents the mixing of hydrogen and oxygen. The electrolytic process is as follows:

$$H_2O \rightarrow H_2 + \frac{1}{2}O_2$$

In the coal gasification method coal is pyrolyzed in a gasifier and is oxidized by steam at a higher temperature and a high pressure. The temperature can be as high as 1500°C, and the pressure can be 30 atmospheres or more. At the high temperature, a smaller amount of impurities is produced, and at the high pressure, less energy is needed to compress hydrogen gas, compared to lower temperature and pressure. The products of the reaction are hydrogen and carbon monoxide:

$$C + H_2O \rightarrow H_2 + CO$$

Of the four methods, steam reforming and partial oxidation have been used for the bulk production of hydrogen. The other two methods usually are not chosen: water electrolysis will be used only when the cost of electricity becomes affordable; and the coal-gasification process produces a trace amount of polynuclear aromatic compounds, which have significant adverse health effects that make the method environmentally unattractive.

Hydrogen manufactured by the steam-reforming and partial-oxidation methods contains impurities such as carbon monoxide, carbon dioxide, methane, carbon, hydrogen sulfide, and water vapor; but these impurities can easily be removed. For instance, carbon monoxide can be converted to carbon dioxide by injecting steam into it to produce additional hydrogen gas:

$$CO + H_2O \rightarrow CO_2 + H_2$$

Carbon dioxide can be removed by absorption (Girbotol method) using an amine to hydrogenate the CO_2 to methane. Methane can be removed by refrigeration or by activated carbon adsorption. Carbon can be scrubbed out by water or filtered by various media. Hydrogen sulfide can be removed by using a caustic scrubber or a carbon adsorber. Water vapor is condensed and separated from the hydrogen stream.

The purified hydrogen can be stored in gaseous, liquefied, or solid form. The storage methods differ for each form, depending upon the application and the distance from the manufacturing site to the point of application, as described below.

Gaseous hydrogen is stored either in low-pressure gas holders, high-pressure storage tanks, or refrigerated (cryogenic) storage tanks. These containers are provided mainly at manufacturing sites.

Liquid hydrogen is stored mainly in insulated storage tanks. The insulation usually is cryogenic and consists of a vacuum jacket and multilayer radiation shield, which can keep the temperature of the liquid hydrogen as low as −250°C. Because one tank of liquid hydrogen can evolve into approximately 800 tanks of hydrogen gas, the liquid hydrogen can be transported for remote applications.

In solid form, hydrogen is stored with metals, forming metal hydrides. An alloy, such as iron–titanium ($FeTiH_2$), magnesium–nickel (Mg_2NiH_4), or another one, in granular form, is heated to approximately 300° under vacuum to remove all absorbed gases first; then hydrogen is added at high pressure. The alloy absorbs hydrogen and produces heat while forming a metal hydride:

$$FeTi + H_2 \rightarrow FeTiH_2 + Heat$$

The hydride can be stored in a stainless steel tube and used in an engine. Hydrogen can be released from the hydride by supplying heat to the hydride; the hydrogen-release temperature is approximately 80°C or less. Metal hydrides are developed to liberate hydrogen as required by the engine. From the standpoint of safety, the hydrides sound very attractive because the amount of hydrogen released is limited at a given time. The disadvantage of hydrides is that the solid particles need to be removed after hydrogen is released.

Hydrogen has a wide range of uses. In the chemical industry, it can be used in manufacturing hydrochloric acid, ammonia, and methane, to name a few. Also it can be used in the hydrogenation of various petroleums, in the reduction of metallic ores, and in welding.

Hydrogen can be used as a heating fuel to warm homes, and is a potential liquid fuel in aircraft, jet rocket, automobile, and internal combustion engines. Although the heat content of hydrogen is low, approximately one-third that of natural gas, its flame temperature is above 2000°C; so it has quite different characteristics from those of the fuels for conventional internal combustion engines. Hydrogen is a competitive fuel because of its light weight, low NO_x emissions, and absence of objectionable combustion products; these are important characteristics for its use in city buses and truck fleets in urban areas.

Besides the advantage of decreased air pollution, two disadvantages need to be noted:

- Because storage tanks must be kept at a temperature below the boiling point of hydrogen (–253°C), the necessary insulation may result in such an oversized container that the cost of storage may be prohibitive.
- Hydrogen vapor must be vented to avoid ignition of the vapor and the accumulation of an explosive hydrogen–air mixture.

FUEL CELLS

Fuel cells convert the energy of a chemical reaction directly into electrical and thermal energy. The fuel cell reaction almost always involves the combination of hydrogen and oxygen gases to form water. These cells originally were developed to provide power for the Gemini, Apollo, and space shuttle programs. Because of their extremely low emissions, high efficiency, low noise levels, and ability to be constructed in a modular manner, their application gradually has shifted to utility companies and to commercial facilities. Structure, types, comparisons, applications, and environmental effects of fuel cells are discussed below.

Fuel Cell Structure

The basic components of a fuel cell consist of a negative electrode (anode), a positive electrode (cathode), an electrolyte, bipolar end plates, and a current collector. The two electrodes are connected by the electrolyte and an end plate between them. This arrangement creates a compartment with each electrode, the anode and cathode compartments. Fuel is fed in continuously to the anode compartment, and air or an oxidant is supplied to the cathode compartment. The electrons produced by the chemical reaction are collected by the current collector. The cells can be stacked in series; the fuel stream and the air flow can be arranged in a cross-flow mode, and the current flows perpendicularly to the gas flow.

Fuel Cell Types

There are direct and indirect types of fuel cells. In the direct type, hydrogen and oxygen serve as fuels, and are produced in an independent installation. The indirect type, depending upon the electrolyte (a hydrogen-generating unit) used in the fuel cells, can be classified as follows: polymer electrolyte fuel cells (PEFC, 80°C), alkaline fuel cells (AFC, 150°C), phosphoric acid fuel cells (PAFC, H_3PO_4, 200°C), molten carbonate fuel cells (MCFC, 650°C), and solid oxide fuel cells (SOFC, 950°C). (The approximate average operating temperature for each type of fuel cell is shown in parentheses.)

Polymer Electrolyte Fuel Cells

The electrolyte for this fuel cell is a porous membrane made of a fluorinated sulfuric acid polymer, which is an excellent proton conductor. The electrochemical reactions are:

$$\text{Anode reaction: } H_2 \rightarrow 2H^+ + 2e^-$$

$$\text{Cathode reaction: } O_2 + 4H^+ + 2e^- \rightarrow 2H_2O$$

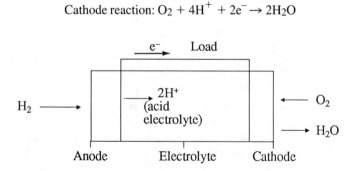

The operating temperature is low because the membrane must be hydrated so that vaporization will occur more slowly than water production. The liquid in the cell is only water; so the corrosion problem is minimal. The thickness of the membrane is determined by its porosity, its wetting property, and the pressure of the fuel gas. The pressure must not be increased so that the membrane is broken or gas comes through the membrane to form an explosive H_2–O_2 mixture.

Alkaline Fuel Cells

The electrolyte is made up of 85 percent by weight potassium hydroxide (KOH) for high-temperature operations (250°C), and 35 to 50 percent KOH for lower-temperature operations ($<$120°C). The electrolyte is retained in an asbestos matrix. The reactions are:

$$\text{Anode reaction: } H_2 + 2OH^- \rightarrow 2H_2O + 2e^-$$

$$\text{Cathode reaction: } O_2 + 2H_2O + 4e^- \rightarrow 4OH^-$$

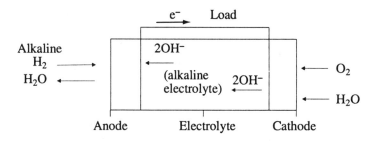

Phosphoric Acid Fuel Cells (H_3PO_4)

Concentrated phosphoric acid is used as the electrolyte, and the fuel cells can be operated at 150 to 220°C. This type of fuel cell is relatively stable, can be operated at elevated temperatures, and has a low water vapor pressure; and the water in the cell is easily managed. The electrolyte is retained in a silicon carbide (SiC) matrix. The reactions are as follows:

$$\text{Anode reaction: } H_2 \rightarrow 2H^+ + 2e^-$$

$$\text{Cathode reaction: } O_2 + 4H^+ + 4e^- \rightarrow 2H_2O$$

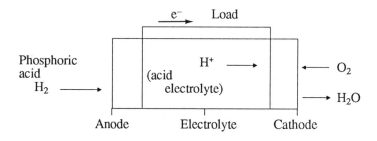

Molten Carbonate Fuel Cells

The electrolyte, which is a combination of lithium, sodium, and potassium carbonates (Li_2CO_3, Na_2CO_3, K_2CO_3), is retained in a ceramic matrix of lithium aluminate ($LiAlO_2$). The fuel cells operate at 600 to 700°C, where the alkali carbonates form molten salt–conducting carbonate ions. The reactions are either with CO or H_2. CO may be obtained by decomposition of methane at a high temperature:

$$2CH_4 \rightarrow 2C + 4H_2$$

$$2C + O_2 \rightarrow 2CO$$

Therefore, the reactions are:

Anode reaction: $CO + CO_3^{-2} \rightarrow 2CO_2 + 2e^-$

$$H_2 + CO_3^{-2} \rightarrow H_2O + CO_2 + 2e^-$$

Cathode reaction: $O_2 + 2CO_2 + 4e^- \rightarrow 2CO_3^{-2}$

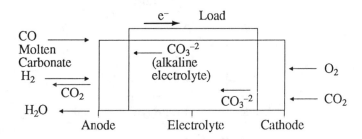

Solid Oxide Fuel Cells

The electrolyte is a solid, zirconium oxide, ZrO_2. The cell operates at 900 to 1000°C, where ionic conduction by oxygen ions takes place. Typically, the anode is a cement made of cobalt-zirconium oxide, $Co\text{-}ZrO_2$, or nickel-zirconium oxide, $Ni\text{-}ZrO_2$; and the cathode usually is $LaMnO_3$, lanthanum manganic oxide. The reactions are:

$$\text{Anode reaction: } H_2 + O^{-2} \rightarrow H_2O + 2e^-$$

$$\text{Cathode reaction: } O_2 + 4e^- \rightarrow 2O^{-2}$$

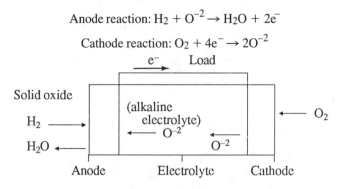

Fuel Cells Comparison

Fuel cells are expected to have a high fuel-to-electricity efficiency, up to 60 percent. Also, because of their modular design, there is some flexibility in the size of fuel cells selected. Most important, the emissions are very low in fuel cell application. A fuel cell's initial cost, however, is very high, and its operational lifetime is relatively short. Also some gaseous reaction products may poison the electrodes. The characteristics of each fuel cell type are summarized below.

PEFC (Polymer Electrolyte Fuel Cell; Proton Exchange FC; Fluorinated Sulfonic Acid Polymer, FSO_2OH)

The polymer electrolyte is nonvolatile, and there is little corrosion. The electrolyte rejects carbon dioxide products; so there may be a buildup of carbonates. With this electrolyte, H_2 is the only suitable fuel for direct oxidation ($H_2 \rightarrow 2H^+ + 2e^-$). It requires high-cost ion-exchange membranes and a thick coating of platinum (Pt) on the electrodes as an electrochemical catalyst, due to the low temperatures needed to achieve a reasonable oxidation rate. Carbon monoxide poisons the catalyst (Pt), and water management may be a problem.

AFC (Alkaline Fuel Cell)

A wide range of elelctrocatalysts can be used. The cost of the catalysts is low, and the hydroxide (OH^-) produced by the cathode can be used by the anode. This fuel cell does not reject carbon dioxide; so carbon-containing fuels are not suitable for it.

PAFC (Phosphoric Acid Fuel Cell)

This is the first fuel cell used commercially. The electrolyte rejects carbon dioxide products, and the fuel cells have a high overall fuel efficiency in on-site cogeneration applications. Hydrogen is the only suitable fuel for direct oxidation, and the water produced at the cathode will not be needed at the anode. The cell uses expensive platinum electrocatalysts, and carbon monoxide is an anode poison. The conductivity of the electrolyte is low.

MCFC (Molten Carbonate Fuel Cell)

The products, carbonates, at the cathode can be supplied directly to the anode. High-grade heat is produced (600 to 700°C), and the cell's efficiency is high. Carbon dioxide sources are required for the reaction at the cathode. The reactants are sensitive to sulfur and to mechanical instability (there is rapid corrosion of metal parts). Long periods of heating are required before useful service is obtained.

SOFC (Solid Oxide Fuel Cell)

The products, oxygen ions (O^{-2}) at the cathode, are supplied directly to the anode. High-grade heat is made available (900–1000°C), and the system's efficiency is high. The electrolyte composition does not change during reaction; so there is no electrolyte management problem. The high temperature requirement limits the availability of cell materials and makes fabrication costly.

Fuel Cell Applications

The technology of fuel cells is being transferred from high-tech applications to commercial and residential uses. The general tendency is for systems sensitive to weight and volume requirements to use pure hydrogen and oxygen as fuels. Space or submarine applications, for example, may use polymer electrolyte fuel cells. On the other hand, systems sensitive to environmental, efficiency, and cost issues should use fossil-fuel-converted hydrocarbon and air as fuels. Molten carbonate fuel cells are the most attractive cells for this application.

Generally, acid electrolytes are suitable for commercial and residential uses because of the manageable low-temperature waste heat. Alkaline electrolytes are used mainly in industrial applications because of the high-quality waste heat generated, which can be used for other processes. Phosphoric acid electrolytes are being successfully used in commercial sites in Japan, Canada, and the United States. Natural gas and steam are mixed to produce hydrogen and carbon dioxide; and hydrogen gas is guided to the fuel cells to react with oxygen to produce steam, thermal energy, and direct electric current, which is transformed to alternating current.

Applications of fuel cells, sensitivity factors, and power ranges are as follows: For space propulsion, weight, volume, and the initial cost are critical; the power range is 15 to 100 KW. For on-site integrated energy systems, reliability and cost are important; a typical size range is 40 to 400 KW. For cogeneration purposes, the overall efficiency of the cogen system and the quality of waste heat are the main considerations in the design stage; the power ranges from 1 to 50 MW. The typical size of fuel cells for an electric power utility is from 5 to 1000 MW; here cost is the major concern. In all these applications, fuel cells offer the excellent advantage that their impact on the environment is minimal.

Environmental Effects

The effects of fuel cells on the environment may be examined from three perspectives: air pollution, thermal pollution, and noise pollution.

It has been reported that with the same power generation rate, fuel cells emit less than 3.5 percent of the NO_x and one hundred thousandth (1×10^{-5}) of the particulate matter than would be emitted from burning natural gas. These emissions are far below those specified in clean fuel burning standards.

Also, it can be shown that fuel cells produce less waste heat than does fossil fuel burning because fuel cell power generation is more efficient than that of conventional power plants. Even if waste heat is generated by fuel cell operation, it normally is ejected into the atmosphere and not released into a body of water; therefore, no thermal pollution occurs.

Moreover, fuel cells react electrochemically; so there are no moving parts to generate noise or vibrations. Although auxiliary devices, such as pumps, blowers, or transformers, are required to complete the whole process, the noise level has been reported to be 50 decibels at 100 feet, a level equivalent to that of a floor fan or humidifier (which is classified as quiet major equipment) in a room.

Therefore, the total impact of fuel cells on the environment is negligible. Some aspects of fuel cells need improvement, however, before they can become mature and competitive products in the marketplace. These needs are discussed below.

Further Developments

The catalysts, the capital cost, and the heat transfer process of fuel cells need further attention.

In a low-temperature operation, precious metals are used as catalysts; but because of high demand and limited quantities, these expensive precious metals may not be available, and fuel cell operations will tend to be high-temperature and high-pressure operations employing cheaper catalysts, such as nickel, nickel oxide, and so on. Better catalysts need to be developed to limit the pollution effects of high temperatures and pressures.

The capital cost of fuel cells is still too high; it was reported that the cost was $2500 per kilowatt for units installed in 1991. The cost should be reduced to one-tenth this value to make fuel cells competitive with other power generators. (In mid-1991 a fuel cell power plant was quietly operating at a 20 MW rate in San Ramon, in Northern California; and a plant with ten times that power rate was being planned for Santa Clara, California.) To protect fuel cell materials from corrosion and to keep electrolytes from being decomposed, the water produced and approximately one-third of the heat generated from the reaction must be disposed of. These problems should be solved in the development of new fuel cells.

SUPERCONDUCTORS

Superconductors have unique properties that have been used in developing advanced energy sources and their high-tech applications. The overall efficiency of the applications is much higher than for those using conventional energy sources, and their environmental effects are negligible. The phenomenon of the superconducting state, the properties of superconductivity, and superconductor applications in environmentally sensitive areas are summarized below.

Superconducting State

Many metals, such as lead, tin, aluminum, niobium–titanium alloy (NbTi), niobium–tin alloy (Nb3Sn), niobium–germanium alloy (Nb3Ge), and so on, lose their electrical resistance sharply when the temperature decreases below a certain point near absolute zero, 0°K. (This phenomenon was discovered by Kamerlingh Onnes in 1911.) The electrical conductivity of the metal thus is very high in this state, suggesting the name "superconducting state." The transition point is known as the transition or critical temperature, T_c. The value of the critical temperature differs; generally pure elements have lower and alloys have higher critical temperatures. The following list gives some examples.

Material	Critical temperature, °K
Aluminum, Al	1.2
Niobium, Nb	9.3
Tin, Sn	3.7
Lead, Pb	7.2
Titanium, Ti	0.4
Niobium–tin alloy, Nb3Sn	18.1
Nb–Al–Ge alloy	23.0

Researchers are looking for materials with high critical temperatures and have found some peculiar properties of metals in the superconducting state. The *Wall Street Journal* (May 24, 1991) reported that buckyballs (spherical molecules of pure carbon) can be made into superconductors when mixed with potassium or rubidium at approximately 30°K (−405°F). The most desirable critical temperature is 77°K (−320°F), the temperature of liquid nitrogen, where the most promising applications of superconductors operate.

Properties of Superconductivity

Electrical Resistance

For practical purposes, the electrical resistance in the superconducting state may be considered to be zero. When a current is induced in a nonsuperconducting coil at low temperatures, it will decay in one second or less. When the current is induced in a superconducting coil, the decay time reportedly is several years. It has been established that the resistance in the superconducting state is at least 1000 times less than that in the normal state.

Magnetic Field Exclusion

When a metal specimen is placed in a moderate magnetic field, a magnetic flux will pass through the interior of the specimen. When the temperature is reduced and cooled down below the metal's critical temperature, condensation energy is released from the specimen. The condensation energy is released in two ways: one portion of the energy is consumed to force the magnetic field out of the interior of the specimen (exclusion); the remaining condensation energy is dissipated into the atmosphere. Thus, magnetic fields are excluded from a superconductor in a superconducting state. (This was first discovered by W. Meissner and R. Ochsenfeld in 1933.) When the intensity of the applied magnetic field is increased to overcome the condensation energy, one portion of or the total magnetic flux penetrates into the interior of the specimen so that the superconductor will be partially or totally in a normal state. The manner in which this phenomenon occurs depends on the type of superconductor.

Type I Superconductors

This type of superconductor has a critical magnetic field (Hc) that increases with a decrease in temperature below the critical temperature. When the applied magnetic field is below the critical magnetic field, the total magnetic flux is excluded from the Type I superconductor. This condition is known as becoming completely diamagnetic. If the applied magnetic field exceeds Hc, the entire superconductor reverts to the normal state, and the magnetic field penetrates into the superconductor completely.

Type II Superconductors

Type II superconductors have two (lower and upper) critical magnetic fields, Hc1 and Hc2. When the applied magnetic field is below Hc1, the magnetic flux is completely excluded. When the applied magnetic field is slightly above Hc1, the magnetic flux begins to penetrate into the superconductor in filaments known as fluxoids. Each fluxoid consists of two parts: a central core and a zero-resistance tube surrounding the core. The magnetic flux is concentrated in the core, and the zero-resistance supercurrent flows along the tube. This condition is known as the mixed state, covering the region between Hc1 and Hc2. When the applied magnetic field exceeds the upper critical field, Hc2, the superconductor performs in the same manner as in a normal state. Because a zero-resistance supercurrent can flow through the mixed state where a magnetic flux exists, Type II superconductors are found to have practical importance in the application of high-field magnets.

Type II superconductors have some loss of energy in the mixed state. When the supercurrent flows along the zero-resistance tube, a force is created to move the magnetic core. The movement of the magnetic core associated with electromagnetic induction creates voltage, and in the presence of a current, energy is dissipated. The fluxoids can be prevented from moving by introducing an impurity in the specimen to minimize energy dissipation.

Thermal Conductivity

In a normal state, the thermal conductivity and the electrical conductivity of a metal, such as copper wiring, are coupled together. In a superconducting state, the situation is completely different. A pure superconductor has less thermal conductivity in the superconducting state than in the normal state. Especially as the temperature, T, approaches $0°K$, the thermal conductivity vanishes in the superconducting state although the electrical conductivity becomes almost infinite.

Generally, heat is an indicator of the degree of disorder (or entropy). The third law of thermodynamics states that the entropy of any system will reach the same least value for every state of least energy as the temperature reaches zero ($T \rightarrow 0$). This implies that the superconducting state is in a quasiperfect order, and that no disorder exists to indicate heat. Therefore, no thermal conductivity is present in the superconducting state.

This property has been used to build heat switches at low temperature. A superconductor wire is connected to two bodies; and by applying magnetic fields, the superconducting state may turn on or off to disconnect or connect the thermal contact of the two bodies.

Other Physical Properties

Although electrical, magnetic field, and thermal properties change in the superconducting state, the mechanical, elastic, density, and tensile strength properties, among other physical characteristics, do not change, or change very little. Therefore, superconductors can be used in large-scale applications for energy conservation with minimal environmental problems. Their applications to electric power utility systems and transportation systems are discussed below.

Superconductor Applications

Electrical Utility Systems

The application of superconductors in an electrical utility system can be divided into power generation and energy storage. Superconductors also can be used in connection with both conventional and advanced technology to generate electrical power efficiently and economically. Their special features are indicated below.

Power Generation

Small-scale superconducting a.c. generators have been successfully demonstrated for years. The typical superconducting generator consists of two parts, a rotor and a stator. Magnetic fields provided by the superconducting coils in the rotor cut the stator windings (armature) to produce electricity. The advantages of superconducting power generation are their negligible electrical loss, small size, low weight, low installation costs, low machine interference, stability, and high-voltage feasibility.

Superconducting magnets can be used to surround a tube of plasma—a combination of neutral atoms, positively charged ions, and electrons in a gaseous state. The magnetic fields cause positively charged ions to turn into one wall of the tube and push the electrons into the opposite wall. Charge collectors along the walls gather electrical current directly from the hot gas. This is superconducting magnetohydrodynamic generation; no moving parts are involved. Electrical current can be obtained from the hot gas, and the thermal energy of the hot gas may be recovered at boilers for steam power generation.

In the future, superconducting magnetic fields may be used to control fusion energy—energy released from the joining of the nuclei of two hydrogen atoms, or that of a deuterium and of a tritium, forming a helium atom, or a helium atom and a proton, respectively (Chapter 10).

The generation of electrical power is but the first step in the electrical utility system. Energy storage, described below, is as important as power generation.

Energy Storage

The excess electrical energy generated by a power plant can be stored in superconducting coils, and the stored energy can be drawn back to the power plant when extra energy is needed. The superconducting coil energy storage system is an electrical circuit that contains kinetic and potential energy and has the property of inertia. The kinetic energy appears in current form as electrons circulating in the coil. The potential energy is stored as charge in capacitors and as magnetic fields in the coils. The property of inertia means that it is initially hard to establish a current in a large-scale coil, but once it is created, it is hard to stop it. Stored energy will heat up ordinary wire, and may burn off the circuit. Superconductors have no resistance, so no burnoff will occur; therefore, superconductors can be used for long-term energy storage.

All energy stored in superconducting magnetic fields can be used by controlling the current circulating in the coils. The energy can be supplied for the following purposes:

- The extra energy required at peak demand times.
- An instant power supply to allow time for standby power generators to start up.
- Absorption of excess power for later use.
- Stabilization of voltage.
- Supplying enormous amounts of energy rapidly.

Transportation Systems

Superconductors have been investigated for transportation applications: on land, in the magnetic levitated train, and on the sea, in the marine propulsion motor.

Magnetic Levitated Trains

A magnetic levitated train is a new design using innovative concepts to lift the train up from its track and then to propel the train along the track. The magnetic force can lift an entire train approximately one foot above the rail, floating it in the air. Its traveling speed can reach 300 mph. Two types of levitation are under development: attractive and repulsive levitation designs.

The attractive levitation type is based on an iron T-beam and superconducting magnetic coils. The T-beam is elevated from the center of a railroad bed, and the train embraces the T-beam from the top around both sides of the beam. The superconducting magnetic coils are mounted upon the lower legs of the train. When the current is charged to the superconducting coils, the superconducting magnet can pull the train upward to the T-beam. This method is simple and inexpensive; however, the levitation force can be so strong that the train is stopped because of grinding at the contact surface of the legs against the T-beam. Constant adjustment of the superconducting magnetic current, and thus the train elevation, is necessary to avoid contact between the train and the T-beam. This, in turn, makes the vehicle unstable.

The repulsive levitation type is the opposite of the attractive levitation design. Basically only two elements are involved: an ordinary loop of wire without magnets and a set of superconducting coils. The ordinary loop of wire is placed on a railroad bed, and the superconducting coils are mounted on the train. When the magnet-laden train runs over the loop, large currents are induced in it, generating a magnetic field that opposes the train. The induced electromagnetic force is the repulsive force that levitates the train into the air.

A possible problem with repulsive levitation is that the current induced in the loop remains for only a fraction of a second so that the induced magnetic field diminishes in a very short period of time. Therefore, the train can be lifted only for a short period of time—which suggests that repulsive levitation is effective only when the train travels at a high speed. However, the problem can be solved simply by using conventional wheels in the period of slow speed.

A complete levitation system will include both attractive and repulsive levitation techniques, using liquid helium—possibly with liquid nitrogen in the future—as a coolant to maintain the superconducting state.

After the levitation of a train, it also can be propelled by superconductivity, and the driving force can be delivered by a system known as a linear synchronous motor. The motor consists of a series of wires set in the railroad bed, superconducting magnets, and a magnetic field reversing system on the train. When the train is approaching a set of the railroad bed wires, current is charged to the wires, generating a magnetic field that attracts the train's superconducting magnets and thus pulls the train forward. When the train passes over the wire set, the train magnet is reversed so that the same track's magnetic field repulses the train magnets and pushes from the rear forward.

The above system is called a linear synchronous motor because it is a motor with linearly stretched wire, and the track current is synchronized with the movement of the train. With this installation, the speed of a levitated train can be controlled by setting the frequency of the propulsive magnetic wave.

The magnetic levitated train will become more competitive in the land transportation system when high-temperature superconductors are available.

Marine Propulsion Motors

The principle of using superconductivity to generate electricity is being used to build propulsion motors for marine vessels. The generated electricity is conducted to an electrical motor that is directly connected to propellers forming a "pot." Because of the use of superconductivity, the size and the weight of the superconducting power generator are reduced significantly, compared with the conventional nuclear, diesel, or steam engine. Therefore, no weight balance consideration is required. Also the superconducting power generator and the pot do not need to be placed closed to each other; they do not even have to be aligned in a straight line as a mechanical system must. Thus the pot can be located at an optimal place where it is hard to detect the propeller noise; no hull profile restriction applies.

The drawback of superconductors is the high cost of the refrigeration needed. However, once the high-temperature superconducting materials are developed, the cost should not be as critical as it is now.

Appendix A:

Control of Particulate Emission

INTRODUCTION

This section begins with a brief discussion of the history of air pollution in order to classify air pollution problems. The main focus of the appendix is on the industrial control of particulate emission, from an engineering point of view.

It was not until the twentieth century that the resources of science and technology began to focus on municipal air pollution. Engineers, meteorologists, chemists, and physicists began to investigate air pollution, either because of contamination in their own work areas or to attack specific problems directly. Consequently, considerable knowledge about the nature and control of air pollution emerged.

Once fossilized energy was released in the form of petroleum and natural gas, it not only transformed industrial and domestic heating practices, but changed transportation almost completely, leading to the use of gasoline or oil for internal combustion engines, replacing earlier modes of power for highway transportation, railways, marine transport, and airplanes. As a result, the combustion residues of petroleum products have contributed to community air pollution.

The problem of air pollution, commonly referred to as "smog," has been addressed by many professional disciplines, including engineering, law, public administration, economics, medicine, and most of the major fields of pure and applied science. Although research findings cannot be said to be scientifically complete, the control of this historic problem is now within reach. (It should be noted that the term "smog" originally was derived from the English word "smoke-fog," but now means a complex mixture of smoke, dust, fumes, gases, and other solid and liquid particles.)

The Air Pollution Problem

Actually, no area in the world has ever had completely pure air; some natural pollution—windblown dust, smoke from forest fires, salt particles from the ocean—has always been present. The test for good air quality is not whether the air is pure, but whether it contains substances harmful to plant or animal life. Nature keeps the air fairly clean; pollutants are dispersed by wind, some are washed to the ground by rain, and others are absorbed by plants. It is only when the pollution effect of human activities overtakes nature's ability to purify that major air pollution problems develop.

Many writers have maintained that air pollution nuisances have spanned thousands of years. The various periods of human history—the fire, copper, bronze, iron, and atomic ages—offer evidence that humans have always engaged in air-polluting activities. Initially, these events were isolated in small settlements and towns, and were treated as individual cases of smoke and fume pollution, affecting only those persons living close to their sources.

In the early periods of human history, wood was the prime source of energy; its use slowed the evolution of industrial processes and limited the availability of heat energy. However, with the impact of the industrial revolution in the eighteenth and nineteenth centuries, cities grew, and air pollution nuisances increased in frequency and complexity, involving entire metropolitan air spaces.

Photochemical Oxidants

In 1952 Dr. A. J. Haagen-Smit, known as the "father" of smog control, discovered the photochemical process that created smog in Los Angeles: hydrocarbon vapor from petroleum refining and automobile exhausts were reacting with oxides of nitrogen in the presence of sunlight. He also demonstrated that eye irritation, damage to green leaves, and light-scattering irradiation were caused by a mixture of hydrocarbon vapors and nitrogen dioxide.

Most nitrogen oxides in the air are emitted as nitric oxide, which is oxidized rapidly to nitrogen dioxide. This NO_2 then reacts with certain hydrocarbons, using energy from sunlight, to form photochemical oxidants. A photochemical oxidant includes several different pollutants, but primarily consists of ozone plus a group of chemicals called organic peroxynitrates that comprise a very small percentage of the total oxidant. Ozone, which is the oxidant in what people commonly refer to as "smog," is a colorless toxic gas that may affect biological systems at extremely low concentrations.

To prevent toxic accumulation of ozone, it is necessary to control the sources of both hydrocarbon vapor and nitrogen oxides, which probably would be harmless except for the toxic compounds they engender.

Air Contaminants

Contaminants, or pollutants, usually are considered to be only those substances that, added in sufficient concentration to the atmosphere, produce a measurable effect on human beings or other animals, vegetation, or material. They may include any natural or artificial composition of matter and may occur as solid particles, liquid droplets, or gases, or in various mixtures of these forms. The types and sources of these contaminants are given below.

Solid Particulates

Source	Solid Particulate Matter
Wood burning	Fly ash
Combustion of fuels	Carbon, soot particles
Metal industries, refineries	Metallic fumes, metal oxides
Mineral industry (stone, coal, salt)	Mineral dusts

Liquid Particulates

Source	Liquid Particulate Matter
Battery manufacturing, fuels combustion	Acid droplets
Petroleum refining, asphalt paving, roofing	Oily and tarry droplets
Food can industry, automobile industry	Paints, surface coating droplets

Gaseous Pollutants

Organic Gas

An organic gas is any chemical compound that contains carbon except carbon monoxide, carbon dioxide, carbides, and carbonate compounds. This group of gases consists entirely of compounds of carbon and hydrogen and their derivatives, and includes all classes of hydrocarbons (olefins, paraffins, and aromatics) and the compounds formed when some of the hydrogen in the original compounds is replaced by oxygen or other substituents. The principal source of hydrocarbons is petroleum, and the major sources of emissions of hydrocarbons and their derivatives are those related to the processing and use of petroleum and its products. (More details are presented in Appendix B.)

Inorganic Gases

An inorganic gas is any chemical compound that is composed of matter other than plant or animal substances. This includes minerals, oxides of nitrogen, oxides of sulfur, carbon monoxide, ammonia, hydrogen sulfide, chlorides, and many other compounds. The principal source of gaseous pollutants is the combustion of fuel for industrial, commercial, and domestic uses. Other gases are emitted in connection with certain industrial processes. (Control of these emissions is discussed in Appendix B.)

Pollution Sources
Generally sources of pollution are classified into two categories: mobile sources and stationary sources.

Mobile Sources
Mobile sources include all transportation carriers, such as airplanes, ships, and motor vehicles; these sources move from place to place, county to county, and state to state. Control of mobile sources is the responsibility of the state and federal governments. For example, the agency for the state of California is the California Air Resources Board, or ARB, established in 1969 with its main headquarters in Sacramento. The primary agency for the federal government is the U.S. Environmental Protection Agency, or EPA, with its main headquarters in Washington, D.C. The United States is divided into ten regions, all of which are under the jurisdiction of the EPA.

Stationary Sources
Stationary sources include agricultural, commercial, domestic, and industrial sources. The control of air pollution from these stationary sources is the primary responsibility of the local air pollution control district.

Agricultural sources result from the burning of agricultural and wood wastes. They are known as area sources, or non-point sources. In October 1976, it was reported in the publication *Point O Eight* that the control of non-point sources of pollution, including agricultural and urban runoff, was just beginning. "Point O Eight" means 0.08 part per million, which was the federal air quality standard for an oxidant (ozone), later revised to 0.12 ppm on January 26, 1979.

Commercial Sources
The commercial sources category includes vapor recovery systems, storage tanks for gas stations, solvents, dry cleaners, smoke from ovens in restaurants, and other sources.

Domestic Sources
Domestic sources are those sources related to the household or the family, such as cooking facilities; they include barbecues and heating systems and all household equipment. The control of equipment used for housekeeping purposes

in households of not more than four families is exempt by law from government regulation.

Industrial Sources

Control of particulate emissions from industrial sources is the main subject of this appendix. The principles of control, properties of particulates, and important elements of individual control equipment are discussed following the summary of California and federal ambient air quality standards given below.

In 1959, the California State Department of Public Health published the following statement in Section 426.1 of the Health and Safety Code:

> The State Department of Public Health shall, before February 1, 1960, develop and publish standards for the quality of the air of this state. The standards shall be so developed as to reflect the relationship between the intensity and composition of air pollution and the health, illness, including irritation to the senses, and death of human beings, as well as damage to vegetation and interference with visibility.
>
> The standards shall be developed after the department has held public hearings and afforded an opportunity for all interested persons to appear and file statements or be heard. The department shall publish such notice of the hearings as it determines to be reasonably necessary.
>
> The department, after notice and hearing, may revise the standards, and shall publish the revised standards, from time to time.

The following is a simplified table for California and Federal Ambient Air Quality Standards.

California and Federal Ambient Air Quality Standards

Pollutant	Averaging time	California standard	Federal standard*
Photochemical oxidants	1-hr avg.	0.10 ppm	0.12 ppm
Carbon monoxide	1-hr avg.	20 ppm	35 ppm
Nitrogen dioxide	Annual average	0.25 ppm (1-hr avg.)	0.05 ppm
Sulfur dioxide	24-hr avg.	0.25 ppm (1-hr avg.)	0.14 ppm
Suspended particulate matter	24-hr avg.	50 $\mu g/m^3$	150 $\mu g/m^3$
Lead	30-day avg.	1.5 $\mu g/m^3$	1.5 $\mu g/m^3$ (calendar quarter)

*National primary standards: the levels of air quality necessary, with an adequate margin of safety, to protect the public health. Each state must attain the primary standards no later than the due date specified in the 1990 Amendments to the Federal Clean Air Act. The above table reflects the air quality standards of 1990.

These standards have been used by individual agencies to develop a control strategy. In addition to the right selection of control equipment, optimal sizing of the equipment plays an important role. The ability to make the correct judgments and to size equipment optimally depends on both technical knowledge and experience.

PRINCIPLES OF PARTICULATE EMISSION CONTROL

Particulate matter is defined as any type of material, such as smoke, dust, fumes, mists, or sprays, that exists as a solid or a liquid under standard conditions. Those particles with a diameter of less than 10 microns are called PM10. The principles applied to control these particulates are one or a combination of the following mechanisms: force of gravity, momentum separation, centrifugal force, inertia impaction, interception, diffusion, and electrostatic forces.

Force of Gravity

The simplest method of removing particulates from a moving gas stream is to allow them to settle out under the force of gravity. Large particles often settle on the floor of a horizontal duct as if it were designed as a simple settling chamber. This settling method usually is applied to the larger or coarser particles, defined as those unable to pass through a 200 mesh screen, or a screen larger than 76 microns.

Momentum Separation

The momentum separation method depends on producing a sudden change of direction in a gas stream. The particles, because of their inertia, will continue to move in the same direction as the initial gas flow and move into a collecting hopper while the gas stream leaves the collector. Momentum separators are slightly more complex in construction than settling chambers, but take up less space.

Centrifugal Force

Centrifugal force is the basis for the most common dust removal method in the industry. With this method, particles are separated from spinning gases by centrifugal force. In comparison with gravity force and momentum separation methods, the following observations can be made:

- Because centrifugal force is greater than the force of gravity, it is more effective in removing smaller particles than are gravitational settling chambers, and requires much less space to handle the same gas volume; on the other hand, the pressure drop is greater, and power consumption is much higher than with gravitational settling.

- The distinction between centrifugal and momentum separators is that the momentum separator is simply a diversion of the gas stream from its original path, whereas in the centrifugal separators there are a number of revolutions of the gas flow.
- Gravity settling chambers and momentum separators are used to collect coarse grit (particles larger than 75 microns), whereas centrifugal separators are effective in collecting particles down to 10 microns in diameter.
- For very hard and large grit particles, momentum separators or gravity settling chambers are used because of erosion of the centrifugal separator wall.

Inertia Impaction

When a dust-laden gas impinges on a body, the gas will be deflected around the body while the dust particles tend to be collected on the surface of the body. In general, impingement collectors are designed for a pressure drop in the range of 0.1 to 0.5 in. H_2O, and limited to removal of dust with a diameter larger than 10 to 20 microns. There are several considerations to be aware of:

- A rapper is needed to shake the collected dust off the collecting body.
- The adaptability of the inertia impactor to the existing flue work is a plus.
- The inertia impactor can be used at a high temperature but not if the dust becomes sticky.

Interception

If a particle is large enough, and the particle center lies closer than one-half of its diameter to the collector, the particle will touch the collector and be intercepted. This mechanism has been used in the following equipment:

- Water spray tower
- Venturi scrubber
- Baghouse

Diffusion

Many small particles in the submicron range not only follow the streamlines but also move across them in an irregular way. This zigzag movement of the small particles, caused by the rapid movement of molecules of gas, is called Brownian motion. In a still gas, small particles move freely and distribute themselves evenly throughout the gas; if a large object were placed in the gas, some of the particles would settle on it and thereby be removed from the gas. Because the time for the process of dust removal by diffusion is limited, the gas in the streamline should remain sufficiently close to the collector. A baghouse (to be discussed later) is an example of equipment that uses diffusion to collect dust.

Electrostatic Forces

Electrical forces can be present on either the particles, the collectors, or both. Four aspects of electrical forces acting in a system of particles approaching a collector should be considered:

1. When *both* the particles and the collector *are charged,* Coulombic forces of attraction or repulsion act, depending on whether the particles and the collector have different or the same charges. These are considered as point charges (Coulombic force).
2. A *charged collector* includes a charge on the particle surface, opposite in sign to the charge on the collector. This force is an additional force on the particle.
3. If a *particle is charged,* it, in turn, induces an image charge of opposite sign on the collector. This results in a force that is an additional force between the particles and the collector.
4. Because the particles are *charged in the same sense,* a repulsive force will be produced among them. This is called the *space charged effect.*

The force between a charged aerosol particle and a charged collector with constant charge is given by the sum of these four forces.

PROPERTIES OF PARTICULATE MATTER

Although initial cost, operation and maintenance costs, space, arrangement, and construction material are the significant considerations in the selection of control equipment, knowledge of the properties of particulate matter is important to obtain an optimum device.

Particle Size

Particle size (diameter) is measured in microns, where 1 micron = 0.001 mm. The mesh used for designation of the screen size is the number of openings per linear inch. The efficiency of control equipment is a function of particle size.

Particle Shape

In practice, the shape of particles is not spherical; surfaces of contacting bodies contain roughness, causing changes of the area of contact, and, hence, the adhesion of particles. The shape of a particle may be taken into account by means of a sphericity factor, S, which has the following values for the shapes indicated:

Spherical: 1.0
Isometric: 0.9
Rounded: 0.78
Soil: 0.67
Elongated prismatic: 0.59
Plane in form of sheet or shell: 0.42

As the sphericity factor rises, the adhesive forces diminish, owing to a reduction in the contact surface of the particles.

Particle Density

The centrifugal force affecting the particle is proportional to the density of the particle: the higher the density, the larger the centrifugal force. Therefore, high-density particles are more easily collected than the lighter particles.

Agglomeration Tendency

This is seen in the tendency of dust particles to bond together loosely to form larger particulates. Generally, after the addition of heat or pressure, the surfaces of particulates may be conditioned to develop adhesive characteristics, which cause them to gather themselves into balls.

Corrosiveness

The occurrence and the rate of corrosion are influenced by the temperature, velocity, pH (0: acidic; 7: neutral; 14: basic), oxidation state, and moisture. The rate of corrosion and its amount depend upon specific conditions. For example, stainless steel does not corrode under any conditions, including an acidic environment. Ferrous metals will corrode when pH $= 0$, but not at pH $= 14$. Therefore, under acidic conditions, stainless steel usage is preferred. Under basic conditions, the less expensive ferrous metal is the metal of choice. In a neutral environment, a nonferrous metal, such as copper or nickel, is frequently used.

Hydroscopic Tendency

The tendency of dust to have a variable moisture content affects the agglomeration of the dust and the corrosion rate of the dust collection equipment. Generally, very humid dusts are the most corrosive.

Stickiness

With sticky particles, special care must be taken in gas distribution, especially immediately upstream of the collecting equipment. If a gas distribution device is selected improperly, sticky dust will adhere to the device, clogging the openings and hindering the flow. The gas flow will be poorly distributed, causing a drastic drop in dust control efficiency.

Inflammability

If the particulate is highly inflammable, the gas mixture can be ignited in its ignition range. A decrease in temperature, pressure, and ductwork size (which affects the flow velocity) will reduce the range of ignition of the flammable mixture.

Toxicity

Toxicity may occur with either inorganic or organic mist.

Inorganic Mist
Inorganic mist is chemical droplets floating in the atmosphere, such as those of hydrochloric, sulfuric, nitric, and chlorosulfonic acids. The mist is highly toxic. For example, sulfuric acid (H_2SO_4) mist is an inorganic mist that has been known to cause death among pigs at a relatively low concentration (5 ppm) and for short exposures (2 to 3 hours). It was reported that the H_2SO_4 mist, starting at a level of 2.5 ppm and 2 to 2.5 microns mass median diameter (MMD), caused the pigs to suffocate.

Organic Mist
An organic mist is mainly hydrocarbon fine droplets. Some mists are less serious, but others are toxic. For example, oil mists belong to this category. Vegetable oils have no pathological effects; peanut oil and corn oil, once accumulated in the lungs, will decrease in concentration progressively; but the quantity of mineral oil and motor oil in the lungs will remain unchanged. It was reported that following heavy exposures of a few grams per cubic meter for a period of 2 to 4 weeks, mineral oil and motor oil resulted in pneumonia. Usually, at below-normal temperatures the toxicity of organic mists will decrease. In these

circumstances, a special acid mist precipitator or an afterburner should be selected for their removal.

Electrical Conductivity

Electrical conductivity is an important factor in choosing equipment and techniques for dust collection. When combustion occurs with a high sulfur-containing coal, the resulting flue gas has a high concentration of sulfur trioxide (SO_3). Because this gas is conductive, it can easily be collected on an electrically charged collecting plate. If low SO_3-containing particles cover the plate, no further collection is possible until these particles are removed.

With these properties of particulate matter in mind, a few selected types of control equipment are discussed in the following sections.

CENTRIFUGAL SEPARATORS

Centrifugal separators consist mainly of three types: simple cyclone separators, multiple cyclone separators, and mechanical centrifugal separators.

Simple Cyclone Separators

Simple cyclone separators consist of a cylindrical tube with or without conical chambers in one or more stages, with gas entering either tangentially or axially. From the top view of a simple cyclone separator with tangential inlet, it can be seen that the gas flow will circulate along the cylinder wall. Particles will tend to move toward the outside wall because of their centrifugal force and may go into a hopper. Clean air will leave the tube, usually through an upward-facing central opening. For the axial flow entrance type, the gas flow will be rotated by means of swirl vanes at the cyclone inlet. Dust will be moved toward the wall and collected at the hopper; clean air will flow upward to the center exit.

Size

There are standard proportions for cyclone design, but there is no generally accepted formula for predicting cyclone performance. The most satisfactory expression for cyclone performance is the empirical one of Rosin, Rammler, and Intelmann (*Z. Ver. Deut. Ing.,* 76, 433–437, 1932), which was plotted in the form of a curve (not shown here) relating collection efficiency to particulate size. The curve was based on experimental data for the variety of dust studied.

Design Criteria

The best cyclones are designed to meet specified pressure drop limitations with the highest efficiency. For ordinary installations, operating at approximately atmospheric pressure, fan limitations generally dictate a maximum allowable pressure drop corresponding to a cyclone inlet velocity in the range of 20 to 70 ft/sec. Consequently, cyclones usually are designed for an inlet velocity of 50 ft/sec although this need not be true.

The primary design factor that can be utilized to control collection efficiency is the cyclone diameter (the smaller-diameter unit operating at a fixed pressure drop having the higher efficiency). However, a small-diameter cyclone will require a lot of units in parallel for a specified gas volume capacity, and the structure will be complicated. The final design involves a compromise between collection efficiency and equipment complexity.

It is usual to design a single cyclone for a given capacity, going to multiple parallel units only if the predicted efficiency is inadequate for the single unit. Cyclones may be used in series when the dust being collected has a broad range of size distributions. In this case, a large-diameter cyclone is used to collect the larger particles, reducing the load and sending smaller particles to smaller-diameter multiple cyclones with high centrifugal force. Multiple cyclones also may be used when the dust in the equipment preceding the cyclone, as well as in the cyclone itself, is poorly distributed (poor distribution usually is the major cause of inconsistent data).

The proportions of the configuration affect the efficiency in certain ways. Reducing the gas outlet duct diameter will improve the efficiency of the collection, as will increasing the length of a cyclone; but the inlet width should be minimized for a given cyclone inlet velocity.

The greatest single cause for poor cyclone performance is the leakage of air into the dust outlet. At this point, the leak can result in a tremendous drop in collection efficiency, particularly for fine dusts; this will create a local nuisance when the cyclone is operated under positive pressure.

Wall roughness does have some effect on pressure drop across a cyclone. Friction causes a reduction of vortex intensity; therefore, pressure drop decreases with the increase of friction. The following results of Rosin et al. show this relationship. With a smooth wall, the cyclone pressure drop indicated as velocity head at the inlet of the cyclone is approximately 8 inches of water. When a light layer of sand particles of small size (approximately 150 to 175 microns) is coated on the wall, the pressure drop is decreased to approximately 6 inches of water; for a heavy coating of the small size particles, to approximately 5 inches of water, and for the large size particles (500 to 1000 microns), to approximately 4 inches of water.

Multiple-Cyclone Separators

A multiple-cyclone separator consists of a number of small-diameter cyclones that operate in parallel, having a common gas inlet and outlet. The gas flow, instead of entering at the side to initiate the swirling action, enters at the top of the collecting tube and has a swirling action exerted on it by a stationary vane positioned in its path. The diameters of the collecting tubes usually range from one foot to as little as two inches. Properly designed units can be constructed to have a collection efficiency as high as 90 percent for particulates in the 5- to 10-micron range.

Mechanical Centrifugal Separators

A mechanical centrifugal separator serves as both an exhaust fan and a dust collector. In operation, the rotating fan blade exerts a large centrifugal force on the particulates, ejecting them from the tip of the blades to a bypass leading into a dust hopper. Its efficiency is somewhat higher than that obtained with simple cyclones. Mechanical centrifugal separators are compact and are particularly useful where a large number of individual collectors are required. This device cannot be used to collect a cake or sticky dust because of accumulation on the rotor blades, which causes clogging and unbalancing of the blades, resulting in high maintenance costs and frequent shutdowns.

WET-TYPE COLLECTORS

General wet-type collectors use a variety of methods to remove contaminant particles from the gas stream. A constant pressure drop and the absence of dust reentrance, as well as the ability to handle high-temperature or moisture-laden gases, give this model an advantage over others. Its other useful qualities include a small space requirement and the ability to treat corrosive gases or dusts, but corrosion-resistant material may add substantially to its cost. Also, the disposal or the clarification of waste water may be difficult and expensive.

For the collection of dusts and fumes, a baghouse (discussed in the following section) is preferred to a scrubber (a wet-type collector). The baghouse ensures complete collection of almost any dust or fume. However, if mists or hygroscopic (moist) particles are present, a baghouse cannot be used. In many cases, a scrubber is the only choice. Mists can be collected successfully in an electrostatic precipitator (discussed in the final section of this appendix). However, if sticky materials are formed, removing the collected material is very difficult, and an electrostatic precipitator is impractical.

Theory of Collection

There are two principal mechanisms by which liquids may be used to remove aerosols from gas streams:

1. Wetting of the particles by direct contact with liquid droplets.
2. Impingement of wetted or unwetted particles on collecting surfaces, followed by their removal from the surfaces by flushing with a liquid.

Mechanisms for Wetting Particles

One mechanism for wetting particles is the impingement of a stream of gas on a liquid spray. This has an efficiency that is proportional to the number of droplets and the force exerted on them. The optimum droplet particle size (diameter) is about 100 microns, according to Johnstone and Roberts; having droplet particles greater than 100 microns allows for too few droplets in the spray, and below 100 microns the droplets do not have enough force. Through diffusion or Brownian movement, dust particles are deposited on the liquid droplets when the droplets are dispersed. This is the principal mechanism used in the collection of submicron particles. A similar method, diffusion due to fluid turbulence, also is used in the deposition of dust particles on droplets.

With dust particles acting as condensation nuclei, gas cooled below the dew point will condense on a wet collector, increasing the particle size and thereby making collection easier. Condensation can be effective only when gases were initially hot. Alone, it can control only a small amount of dust, because removing larger concentrations requires cooling the gases to a greater degree than can be attained.

A process for collecting particles by stimulating their agglomeration has been suggested, through a combination of humidification and electrostatic precipitation. These processes are not well developed, however, and cannot be applied in practice.

A particle may be wetted in two ways: by its making contact with a spray droplet or impinging upon a wetted surface. Once wetted, the collecting process is complete when the particle reaches a collecting surface by impinging on surfaces within the path of the gas flow. The particle may also be collected against the outer walls of the collector if centrifugal forces are applied. The particle may also be allowed to settle by gravity. These mechanisms may be used singly or in combination.

Spray Nozzles

In order to avoid entrainment and act most effectively, the droplets should be sized fairly closely when spray nozzles are used. In addition to this, the nozzle should be able to handle gases with a high solids concentration. There are six nozzle types in use: an impingement type nozzle, a spiral spray nozzle, a spinning disk type nozzle, a two-liquid jet impingement nozzle, a pneumatic nozzle, and a sonic nozzle.

Impingement Type Nozzle
In impingement nozzles, a high-velocity liquid jet is directed at a solid target, which is disintegrated by the impact. Water strikes the metal hood of nozzles

and forms the characteristic spray. [Depending on the type of target, a different (directional or hollow cone) type of spray is produced.] These types of nozzles are not expensive and are simple in construction. They are widely used where coarse sprays are required, as in blast furnace gas spray towers.

Spiral Spray Nozzle

The atomizing action in this type of nozzle is caused by the spinning motion exerted on the liquid before it leaves the nozzle. The liquid enters the spin chamber within the nozzle, passing through the manifold (the outside ring of the chamber) and through a series of tangential holes to the spin chamber (the inside ring of the chamber). The entry holes have a tendency to become blocked by dust; so this nozzle is not recommended when the liquid is being recycled. (Droplet sizes from this nozzle vary between 150 and 400 microns at 50 psi water pressure.) At 400 psi water pressure, a large proportion of droplets of about 100 microns will be produced. This type of nozzle is very effective for centrifugal spray scrubbers.

Spinning Disk Type Nozzle

This type of nozzle is a rotating disk from which liquid droplets are discharged after being accelerated to a high velocity. The flat disk usually is driven mechanically, and the liquid moves across it. The droplets produced are of a uniform size, which can be controlled by changing the disk speed and the flow rate; so these nozzles (or atomizers) are most useful in producing droplets for fundamental studies of their behavior and performance. This technology can be applied for scrubber development where similar-size droplets are required.

Two-Liquid Jet Impingement Nozzle

When two liquid jets impinge, a wave is formed in the liquid at the point of contact. The liquid forms a sheet and then disintegrates immediately to form groups of drops originating from the wave point. The frequency of the waves increases as the jet velocity increases and as the impingement angle decreases, rising to 4000 cycles/sec and forming a large number of fine droplets. These nozzles, or atomizers, have the advantage of producing fine droplet sizes without complex mechanical construction and without wearing down the surfaces of the impingement nozzles. They are not yet used in commercial scrubbers.

Pneumatic Nozzle

In these nozzles, a liquid spray is produced by the impact of a gas stream, usually air, on a liquid jet, rather than by two liquid jets. Pneumatic nozzles produce very fine droplets that have been found in practice to be unsuitable for centrifugal scrubbers because a high-pressure air supply must be provided as well as a liquid (water), adding to the installation and maintenance costs.

Droplet formation in pneumatic sprays has been studied in considerable detail.

Also, determinations of the average drop size and the surface area of the drops per unit volume of gas have been made by researchers.

Sonic (Spray) Nozzle

The liquid flows out of a simple circular chamber through the holes of a nozzle under low pressure (10 psi), and the liquid filaments thus produced are shattered by sound waves at a frequency of 9.4 kilocycles/sec, producing very uniform droplets. Sound waves are produced by the impingement of a jet of compressed air at 15 to 60 psi on a resonator centrally located between the holes. The liquid output from the nozzle is about 6 lb/min.

Types of Wet Collectors

Spray Chambers

A spray chamber is the simplest type of scrubber, consisting of a chamber in which spray nozzles are placed and water is sprayed. The water droplets enter the chamber and contact the particles, which then settle to the bottom of the chamber. In order to prevent liquid from being discharged into the atmosphere, moisture eliminator plates are installed at the outlet of the chamber. Used as a dust collector, the efficiency of a spray chamber is low; however, the spray chamber is extensively used in industry as a gas cooler. Baffle plates are added to improve the dust collecting efficiency, for particles can be impinged upon them. Water spray rates range from 3 to 8 gallons per minute (gpm) per 1,000 cubic feet per minute (cfm) of exhaust air flow.

Cyclone-Type Scrubbers

A simple cyclone type scrubber consists of a dry cyclone with spray nozzles installed. The gas stream flows tangentially to the cylinder, and additional vanes act as impingement and collection surfaces. Helical baffles are installed to enhance centrifugal force, and multiple spray nozzles are used to improve dust wetting. Because centrifugal force is applied in the dust collecting process, the efficiency is increased by relatively high gas velocities. The pressure drop varies from 2 to 8 in. H_2O; water rates vary from 4 to 10 gpm per 1,000 cfm.

Cupola Top Scrubbers

One of the simplest units, this top scrubber often is applied to a cupola furnace. Water runs over a cone-shaped disk, forming an annular water sheet, which is broken up by the furnace gases.

Orifice-Type Scrubbers

In these scrubbers the velocity of the gas is used to promote liquid contact. Air flow is guided through a narrow channel partially filled with water, thus water droplets are dispersed. Centrifugal forces, impingement, and turbulence promote contact between water droplets and their control. The water quantities in motion are relatively large, but most of the water can be recirculated in the scrubbing tank. The recirculation rate is 20 gpm per 1,000 cfm.

Mechanical Scrubbers
Mechanical scrubbers include a rotating drum or disk in the water tank generating the water spray. The turbulence of the spray increases the chances of collision between dust particles and spray droplets. The water usually is recirculated, with rates varying widely among the different types of rotating elements.

Mechanical Centrifugal Scrubbers
A mechanical centrifugal scrubber is a device similar to a centrifugal dust collector—with rotating blades installed inside a container. Water is sprayed to keep the blades wet. When dust particles impinge on the rotating plates, the centrifugal force and water spray flush away the collected dust particles to the outer wall of the container and are then gravity-fed to the hopper, located at the bottom of the container.

High Pressure Spray Scrubbers
The water pressure at spray nozzles determines the intensity of the water spray, which further affects the dust collecting efficiency. A high pressure spray scrubber is operated usually at two to six times higher than normal operating pressure (approximately 100 psi). For multiple spray nozzles, orifices must be placed so as to promote collision between water droplets and dust particles. Water consumption rates range from 5 to 10 gpm per 1,000 cfm of exhaust air flow.

Venturi Scrubbers
As the gas passes through a venturi tube, the pressure at the throat of the tube is reduced. When low-pressure water is added to the throat, fully saturated condensation occurs on the particles in the high-pressure region of the diffuser. This helps the particles to grow, whereas the wet particle surfaces tend to help agglomeration and collection. The wetted particles and droplets are collected in a cyclone-type scrubber, where a very high collection efficiency has been reported for very fine dusts. The water rate is 3 gpm per 1,000 cfm gas, and the water usually is recirculated.

Packed Bed Scrubbers
In packed bed scrubbers, dirty gas is passed through a packed bed of collecting materials—usually coke, rock, or stone contained between support trays. Liquid is passed over the collecting surface to create a thin film and to keep the surface clean. Dust particles are collected as they pass through the packed bed, contacting

the thin film. The packed bed scrubber is not used solely for dust particle removal because as the liquid drains through the packing bed, the collected dust particles may plug the gaps in the packed bed.

Design Criteria and Scrubbing Efficiency

The design criteria for wet scrubbers are high removal efficiency, minimum maintenance, and low capital cost. The collection efficiency of wet collection devices is proportional to the energy input to the devices, and is a function of the energy expended in the air-to-water contact process; higher energy consumption provides more mutual interaction and greater collection efficiency for a given particle size. Because high-energy devices are expensive to install (high capital cost) and operate (high maintenance cost), there is a tendency to install wet collectors of limited efficiency.

As was discussed, the pressure drop is a measure of the energy used and is directly proportional to the power required. For the same dusts, greater scrubbing efficiency requires more power consumption; for finer dusts, the same efficiency requires more power as compared to coarser dusts. Thus the efficiency of collection can be expressed in terms of the total power used and the characteristics of the dust being collected, and is independent of the actual type of scrubber being used.

FABRIC FILTERS

General

In industries, the filtering of particulate matter is of utmost importance, especially in modern society where the air quality has badly deteriorated. The fabric filter is one means of intercepting these contaminants before they are released into the atmosphere. Modern technology has made possible collection efficiencies of up to 99.9 percent. The finer the filtering material, the more effective it is in the filtering process; close knitting of the fabric tends to increase its interference with matter. The disadvantage of using a closely knit fabric is that the resulting pressure loss or power consumption may make it economically unfeasible. An electrostatic charge also can be used in the filtering process. Depending upon the charges and the polarity, an attraction force between the particles and the filter will greatly increase efficiency, whereas a repulsive force will lower the efficiency.

Cleaning

After a filter baghouse has reached its collection limit, the filters are cleaned and made ready for reuse. Baghouses with up to 500 or 600 cubic feet of filtering area usually are cleaned by using a hand-operated lever; a sharp pulse is transmitted to the bag, which shakes the particles loose. Prior to cleaning, the blower must be turned off. Large baghouses use mechanical shaking or reverse air flow cleaning. With a mechanical shaker, a cam converts the circular motion of the motor into an oscillation or a series of pulses, which are directed to the baghouse to separate the dust from the filters.

Reverse air flow cleaning is fully automatic, making use of a programmed cycle initiated at regular intervals or when the pressure reaches some critical value. In this method, one compartment of the baghouse is isolated, and is collapsed by a reverse air flow; and the dust is separated from the filters. After cleaning, the compartment is returned to operation, and the other compartments are cleaned in the same way. The advantage of having a fully automatic system is that the system can be cleaned periodically as it is programmed. However, the possibility of mechanical breakdown may be increased with this process.

Types of Filters

In addition to the regular tubular and cylindrical filters, various other types of filters have been devised. One of these is the multiple-tube bag, where the bag is divided into several separate compartments. Upon inflation of the bag, each compartment attains a circular cross-section, but when the blower is turned off (and the pressure relieved), the bag becomes oval. This shape change aids in breaking up the caked-on particles. This type of filter requires a special and more expensive mounting device than would be needed for a conventional bag with the same filtering area. Its advantage, however, is that for a given amount of space, greater filtering can be achieved than is possible with the conventional design.

Another type is the envelope filter. The bags are mounted on a wire support frame that allows greater filtering area than does a conventional bag for a given volume. The inspection and the maintenance of this type of filter are not so easily accomplished as for conventional bags, and the wear is increased because of frictional forces between the wire frame and the fabric; thus this type of filter should not be used where rapid wear is anticipated.

Filter Media

With the number of materials now available, both synthetic and nonsynthetic, filters utilize different fabrics for different jobs. Factors that determine which type of medium to use range from particle concentration and frequency of cleaning to the temperatures encountered and pressure drops. Listed below are different media and some of their physical properties.

- *Cotton:* Cotton is used as a filter material mainly when the flue gas temperature is low and no abrasive chemicals are involved. It is inexpensive, durable, and the preferred material for many filter applications.
- *Dynel:* Dynel can be used in high-temperature dust filters. It resists burning and chemical abrasion, including the action of acids and alkalis.
- *Nylon:* Nylon is a tough and nonflexing fiber, which is synthetically produced. It resists mechanical and chemical abrasion and has some elasticity. It can be cleaned easily, but its use as a bag filter is limited to very low-temperature applications.
- *Orlon:* Orlon is a lightweight synthetic textile fiber with a high tensile strength, which resists heat and chemical abrasion. It would be a favored material for bag filters, but it is too expensive.
- *Fiberglass:* Fiberglass resists high temperatures and chemical abrasion, but it reacts with fluorine, causing damage to filters. Also it has low resistance to mechanical abrasion and abrupt crushing. Therefore, it should be used only for low-velocity flue gas with gentle shaking.

During the preparation of this manuscript, many new materials have been discovered that have improved heat and chemical resistance and other desirable features. Prospective users should study the current market to obtain the most efficient filtering devices.

Filter Sizing

It is important to understand the effects of pressure drop and permeability in determining the most desirable filter size.

Pressure Drop

Pressure drop is the difference between total pressures at two locations in a gas flow (total pressure being the sum of the static pressure and the dynamic pressure). The pressure drop represents the power required to transport a gas flow from one location to another at a certain rate. If the pressure drop, the flow rate, and the properties of the gas are known, then the horsepower required for the fan can be determined from a fan rating table. The energy or the horsepower required for a fan thus is dependent on the difference between total pressures. The National Association of Fan Manufacturers defines the fan static pressure as the total pressure drop diminished by the dynamic (or velocity) pressure at the fan outlet:

Fan static pressure = Total pressure drop − Dynamic pressure at the outlet

Therefore, the fan tables should be used with care.

Permeability

Permeability is used to characterize the porosity of a filter; it is defined as the flow rate of the air (ft^3/min) that passes through 1 ft^2 of clean filter at 0.5 in. H_2O pressure drop across the filter.

Filter Rate

The filter rate is the velocity of the gas flow passing through a filter bag (ft/min). It was originally defined as the ratio of the gas flow rate (actual cfm) per square foot of filter medium at the actual operating conditions. The filter rate is important in determining the square footage of the cloth required to size the baghouse. The filter rate can be determined only by experience.

Equipment Selection Factors

An attempt was made by Pulverizing Machinery to define the most important factors in proper equipment selection. The results should be considered as a guide, but not as a substitute for good judgment. The factors of interest are: the (particulate) material, application, temperature, fineness, and dust loading factors.

Material

The material factor is a function of the characteristics of the dust to be collected. For physically and chemically stable materials such as sawdust or tobacco, the material factor is large; this means that a relatively small cloth is needed. For an unstable material such as activated carbon or detergent, the material factor is small, indicating that a large cloth is required.

Application

The filtration during an operation should be treated differently from common venting. For example, general venting of a nuisance contaminant requires less cloth than process gas venting.

Temperature

For a gas flow, the viscosity increases with temperature; so more cloth is needed for a higher-temperature gas. For air, the viscosity will increase asymptotically to a constant value, which implies that the result is an asymptotically high temperature factor.

Fineness

The fineness and the characteristics of materials frequently are combined. For a fine material (to be collected), more cloth is needed than for a coarse material.

Dust Loading

With increasing dust loading in a gas flow, the filter rate (which equals the velocity of the gas flowing through the filter) generally will decrease. However, when the gas flow is saturated with dust, it cannot hold more than its saturated limit; so the bag surfaces receive a saturation-limited amount of accumulated dust per unit of time. Thus above a certain dust-loading limit, usually above 100 grains/ft^3, a given bag can handle more material without a decrease in the filter rate.

All five factors are multiplied together to obtain a filter rate. With a given total amount of gas to be treated, at the calculated filter rate, the area of cloth required can be determined:

$$\text{Cloth area} = \text{Total gas flow rate/Filter rate}$$

This filter rate and this cloth area are applicable to high-performance filter baghouses, which use filter media combined with frequent and thorough cleaning (such as pulse jet types and blow ring style units).

ELECTROSTATIC PRECIPITATOR

The control of fine particulate matter, especially particles less than 10 microns (PM10) in size, is a very important health concern because they can easily enter the respiratory system and cause severe health problems. The main sources of these particles are industrial operations, power plants, and commercial entities. Electrostatic precipitators are among the most effective devices now in use for control of PM10.

Aerodynamic, electrical, and mechanical forces are involved in the process of particle collection by means of electrostatic precipitation. Aerodynamic force distributes the exhaust gas, and makes the inlet particles more evenly distributed to the electrostatic precipitator; electrical force pulls particles out of the exhaust gas stream, causing them to adhere to a collecting plate; and mechanical force separates the collected particles from the collecting plate for removal. These three forces are applied directly to particles for removal; therefore, the energy required to operate an electrostatic precipitator is much less than that needed for many other control devices.

The modern electrostatic precipitator consists of internal and external elements. The internal elements include: electrode wires, collecting plates, a rapping system, and a transformer/rectifier set. The external elements include the casing and thermal expansion joints. A brief description and individual functions of each element are presented below.

Electrode Wires

The electrode wires consist of vertically hung wires spanning the full height of the collecting plates. The wires are attached to a frame at the top and held tight by a weight at the bottom, and they are placed between two parallel collecting plates. One wire is placed near the front and another is set near the rear ends of the plates. The space between the plates is usually 12 to 18 inches.

When high voltage is charged to the wires and the collecting plates are at ground level, a large number of positive and negative ions are formed in the vicinity of the wires. This effect is known as a corona, and the wires are referred to as corona electrodes or discharge electrodes. Usually in industrial applications, the discharge electrodes are negatively charged. Electrons are released from the discharge electrodes to ionize gas molecules, creating a corona field; and more electrons are released to create negative ions, which charge dust particles (PM10). The negatively charged dust particles are attracted in the direction of the collecting plates.

Collecting Plates

The collecting plates are called a collector electrode. The collector electrode usually is grounded and attracts the negatively charged particles. In industrial applications, the collecting electrode can be as large as 6 feet deep by 20 feet high or even 8 feet by 25 feet. The edges of each plate must be smoothed out to keep a corona from occurring at lower voltage.

Several plates are mounted on the top frame, on which a mechanical rapping system is installed. The collector plates must be able to stand rapping impacts and transmit the rapping effect evenly over the plate.

Rapping System

The rapping system is used to remove particles adhering to the collecting plates and electrode wires. Impulse (and vibration) methods are used to clean particles from the plates (and wires). The impulse method applies a hammer-impact force to separate particles on a preset schedule, which is controlled by cams mounted on a motor-driven shaft. The cams raise hammers and bump them on the top frame; the collected particles suddenly are released from the collecting plates and fall into collection hoppers.

When the three steps of particle charging, particle collection, and particle removal are performed, the electrostatic precipitation process is complete.

Transformer/Rectifier Set

The energy required to process an electrostatic precipitator must be transformed from 220 volts to 40 to 60 kilovolts. This can be done by using a transformer.

Additionally, the converted high voltage must be changed from alternating current (a.c.) to direct current (d.c.) by means of a silicon rectifier (SCR). The SCR also will give a uniform corona over the entire surface of collecting plates and electrode wires. If the corona is unevenly distributed, the high current may cause sparks and fire in the electrostatic precipitator.

Casing

The above-discussed electrode wires, collecting plates, rapping system, and transformer/rectifier set, which are the internal elements of an electrostatic precipitator, are supported by a casing. The casing is a grounded steel structure in which all the internal elements are suspended to minimize internal stress and external movement. Stainless steel is the preferred casing material to avoid corrosion, and proper heat insulation is needed to prevent damage due to excessively high gas temperatures.

Another thermal effect is elongation and contraction of casing and duct work when the gas temperature varies. This effect can be accommodated by the installation of expansion joints.

Expansion Joints

An expansion joint is specifically designed to absorb different types of duct-work movement induced by thermal expansion. It is installed at the joint between the casing and the duct work as well as at critical locations where deformation due to thermal expansion of the duct work may occur.

Depending upon the location chosen to absorb a specific movement, the expansion joint is designed in different shapes, such as single-bellow, multiple-bellow, or nonmetallic joint types. Sometimes the body of an expansion joint is rigidly connected to duct work to anchor the end; occasionally a double slip joint is used so that both duct work ends are free to move in the joint. The movements of duct work due to thermal expansion are either axial, transverse, angular, or torsional movements.

- For an axial movement, the duct work is under compression and elongation stress. Any type of the above-mentioned expansion joints will suffice for this type of movement.
- For a transverse (lateral) displacement, shear force is dominant, and only nonmetallic expansion joints should be used.
- An angular deflection creates a bending movement that causes the duct work to undergo compression at one side and expansion at the other side. To absorb this strain combination, multiple-bellow or nonmetallic joints should be used.

- For a torsional movement, rotational shear is exerted on the duct work. In this case, only nonmetallic expansion joints should be used.

In addition to absorbing various types of movement, an expansion joint can localize vibration, reduce noise, and compensate tolerances at the duct work. Although the expansion joint is a relatively small part in the whole electrostatic precipitation system, its failure will cause complete shutdown of the entire operating unit. Thus expansion joints play a very important role in electrostatic precipitation systems.

Appendix B

Control of Gaseous Emission

INTRODUCTION

Gaseous wastes from industrial processes are contributing to a rising level of air pollution. These waste gases are pollutants vented from chemical manufacturing processes, the application of coatings, utility companies, petroleum industries, metallurgical industries, and so forth. As mentioned in Appendix A, gaseous pollutants may be separated into two classes, inorganic gases and organic gases, which are described briefly in this section. Health effects of air pollution also are summarized below.

Inorganic Gases

The major pollutants to be discussed here are sulfur compounds, nitrogen compounds, halogen compounds, and carbon monoxide.

Sulfur Compounds

The following are sources of sulfur-containing pollutants: the combustion of coal; petroleum and natural gas—production, refining, and utilities; sulfuric acid and sulfur—manufacturing and applications; and smelting and refining of ores, especially copper, lead, zinc, and nickel.

The Combustion of Coal

The sulfur in coal does not exist in a free state but in combined form, as organic sulfur and inorganic sulfur. Organic sulfur is sulfur combined with carbon in a special form. Inorganic sulfur is sulfur in the pyritic (FeS_2) or sulfate ($CaSO_4$, $FeSO_4$) forms. Organic sulfur cannot be separated by mechanical cleaning processes, but inorganic sulfur can be removed by cleaning operations. The inorganic sulfur content can be reduced by 25 to 35 percent by economical and practical cleaning procedures.

The organic sulfur can be removed by combustion. During the combustion process, a portion of the sulfur is converted to sulfur dioxide (SO_2), which escapes in the flue gases. The conversion of S to SO_2 varies from source to source. The actual sulfur concentration in the flue gases depends on the method of heating, the percent of excess air present, and the sulfur content in the coal. When coal is carbonized to form metallurgical coke, approximately 60 percent or more of the sulfur in the coal remains in the coke. During combustion of the coke, a major portion of the sulfur is converted to SO_2, which escapes through the stack to the atmosphere.

Petroleum and Natural Gas—Production, Refining, and Utilities
In the modern refining process, approximately 40 percent of the sulfur content can be converted to hydrogen sulfide (H_2S). This portion contains enough SO_2 (3.5–7%) that the H_2S can be economically converted to elemental sulfur, which is reusable. The remaining sulfur content needs to be treated by means of pollution-control devices.

Sulfuric Acid and Sulfur—Manufacturing and Applications
The concentration of SO_2 found in the waste gases from sulfuric acid (H_2SO_4) and sulfur manufacture and applications, is relatively low. The pollution problem from these sources is mainly due to sulfuric acid mist emissions rather than SO_2. Different control technologies are available to reduce the sulfur pollutants (see final section of this appendix).

Smelting and Refining of Ores
Because smelter effluent gases contain relatively high concentrations of SO_2 (1–6%), economical recovery of H_2SO_4 and S has been accomplished (Chapter 9).

Nitrogen-Containing Inorganic Gaseous Pollutants
There are two major types of nitrogen compounds in the air, oxides of nitrogen (NO_x) and ammonia (NH_3). Nitrogen oxides are more important than ammonia because NO_x reacts with hydrocarbons under sunlight to form smog, in a process known as the photochemical reaction.

NO_x—Oxides of Nitrogen
The most important sources of NO_x are the exhaust gases of trucks and passenger vehicles and the combustion of natural gas, fuel oil, and coal. Nitrogen oxides also are emitted from the manufacture of nitric acid, sulfuric acid, paint, roofing material, and rubber, and from the refining of petroleum.

Nitrogen monoxide (NO) reacts with the oxygen in air to form nitrogen dioxide (NO_2) gas. Although NO_2 is readily soluble and can be removed, the reaction between NO and O_2 does not go to completion. This allows some NO to be present with NO_2, in a mixture called NO_x.

Ammonia—NH_3

These sources emit ammonia to pollute the atmosphere: fertilizers, organic-chemical industries, the manufacture of nitric acid, and the exhaust of automobiles and refineries.

Some industries spray ammonia in flue gas. Its effect is to balance the acid condition of the flue gas due to SO_2 and NO_x contaminants. However, the reaction of ammonia and acid gases is not always beneficial because it produces an aerosol of fine ammonium salt fumes or droplets, which will make a layer of haze in the sky. Ammonia gas is very soluble in water and can be removed from flue gas by water scrubbing.

Halogen Compounds, Inorganic Gaseous Pollutants

Halogen compounds are nonmetallic chemical compounds and include: fluorine compounds, chlorine compounds and others (bromine, Br: iodine, I; and astatine, At). As far as atmospheric pollutants are concerned, fluorine and chlorine are the most important halogens. Therefore, only they are discussed here.

Fluorine (F) Compounds

Fluorine, which is a nonmetallic element (atomic number 9), is the most reactive element, usually igniting at room temperature. Because of its strong reactivity, it usually exists in gaseous or liquid form. The air pollution due to gaseous fluorides comes mainly from silicon tetrafluoride (SiF_4) and hydrogen fluoride (HF). These two compounds (SiF_4 and HF) can be removed effectively by adsorption, a process discussed later in this appendix. Limestone or water scrubbers also have been used to control these pollutants. Sources of polluting fluorine compounds are aluminum industries, fertilizer manufacturers, steel plants, and refineries.

Chlorine (Cl) Compounds

Chlorine, which also is a nonmetallic element (atomic number 17), occurs as a heavy, noncombustible gas (Cl_2). The source of chlorine compound pollution is refineries, especially at the regenerator of cracking catalysts units. The pollutant usually is emitted in the form of hydrogen chloride (HCl) gas, which is easily recovered by adsorption.

Carbon Monoxide—CO

Carbon monoxide is a colorless and odorless gas, but under certain conditions it also can occur in liquid form. It burns with a violet flame, is slightly soluble in water, is highly toxic when inhaled into the lungs, and is highly explosive in the presence of heat. Major sources of CO are by-products of chemical industries and the combustion of motor fuels with limited amounts of oxygen in trucks and passenger vehicles.

Different means are recommended to control carbon monoxide emissions from motor vehicles, including: engine modification, afterburner-type mufflers, catalytic mufflers, and revised crankcase ventilation systems. The purpose of these methods of control is to convert carbon monoxide to carbon dioxide, which is less harmful than carbon monoxide.

Organic Gases

Organic gases that are considered air pollutants include the following:

- Organic compounds, such as hydrocarbons consisting of the elements carbon and hydrogen, derived principally from petroleum, coal tar, and vegetable matter.
- Chemical compounds, such as mercaptans, that contain sulfur. These compounds are similar to alcohol, with a strong, unpleasant odor.
- Alcohols, such as methanol, ethanol, and so on, which are colorless volatile liquids.
- Organic compounds, such as ketones, containing the CO-group with two hydrocarbon radicals.
- Organic compounds, such as esters, formed by the reaction of an acid and an alcohol.

Sources of organic vapor pollution include:

- Operation of motor vehicles (gasoline and diesel).
- Petroleum refining (evaporation loss).
- Petroleum products (oil wells and natural gas released to the atmosphere).
- Petroleum marketing (gasoline bulk transfer, retail sales).
- Chemical, paint, roofing, and rubber industries.
- Fuel oil and coal burning.
- Incineration of refuse.

Organic vapors are a major source of odors. Types of odors and their sources can be summarized as follows:

- Animal odors: meat-packing, fish oil, poultry ranch, and processing odors.
- Combustion odors: engine exhaust and coke-oven gas odors.
- Food processing odors: coffee and restaurant odors.
- Paint-related industrial odors: paint, lacquer manufacturing, paint spraying, and commercial solvent odors.

- General industrial odors: pulping operation, dry-cleaning, fertilizer plant, asphalt, and plastic manufacturing odors.
- Many others, such as foundry, municipal incinerators, and sewage plant odors.

All these organic emissions can be reduced by add-on control devices, process and material changes, or both.

Health Effects of Air Pollution

The following brief discussion of the health effects of air pollution provides a background for studying individual control technologies.

Typical complaints about air pollution are that it causes annoying odors, irritates eyes or the respiratory tract, reduces visibility, soils paint or clothing, and damages vegetation. These adverse effects usually are detectable, controllable, and/or preventable. Air pollution is considered detectable because meteorological data and simulation can be used to predict its extent and to discover its influence on human life. It is considered controllable because the treatment of air pollution involves reducing emissions through the use of control devices or by alterations either in technology or in the materials used. It is preventable when the pollutants are reduced to meet air-quality standards designed to protect the public health and welfare and to prevent defects caused by the pollution.

Air pollution can be regarded as a disease of modern society, affecting the physical and mental health of the earth's inhabitants. This disease differs from community to community: In London, there are complaints about the black soot produced by burning coal. In California, irritation of the eyes and the throat and damage to vegetation due to photochemical pollution are the main complaints (a situation, as mentioned before, that is due to sunlight irradiating hydrocarbons and nitrogen oxides emitted from vehicles, industries, and commercial and agricultural sources, to form smog).

Not all hydrocarbons and their derivatives (alcohols, ketones, acids) have the same potential for smog formation. The most reactive group of hydrocarbon (the olefins, the unsaturated hydrocarbons) can react with atomic oxygen (formed from the dissociation of NO_2 in sunlight) or with ozone (O_3) to form eye-irritating aerosols, oxidants, and additional ozone. The effects of ozone and other pollutants on human health are summarized below.

Ozone

Ozone is a relatively insoluble gas, but it is highly reactive and has a characteristic odor. Because of its insolubility, most absorption of ozone occurs in the air cells of the lungs. With an increase in the ventilation rate and respiration, more ozone reaches the membrane in the air cells of the lungs, where it may cause a watery swelling and hinder gas diffusion. For some people, ozone causes coughing, headaches, or chest discomfort; so it is advisable to stay indoors, if possible, during ozone episodes, to minimize ozone exposure.

SO_2

Sulfur dioxide (SO_2) is relatively soluble but less reactive than ozone. It too is irritating and has a typical odor. Because of its solubility, during normal breathing most SO_2 absorption occurs in the nose and in the upper airway. Therefore, with an increase in the respiration rate, more sulfur dioxide reaches the windpipe, causing bad coughing. Physical exercise increases the respiration rate and transmits a large portion of the SO_2 to the lower part of the respiratory system. The situation is worsened when sulfur dioxide interacts with particles coming into the respiratory system, as it can aggravate chronic bronchitis.

CO

Carbon monoxide (CO) passes through the respiratory system without modification; most of it is stored in the hemoglobin (an oxygen-carrying substance found in blood) in such a way as to reduce the total number of binding sites available to oxygen. Carbon monoxide binds strongly to hemoglobin, with approximately 250 times the binding strength of oxygen. Therefore, when the ratio of carbon monoxide to oxygen molecules is 1:250, half the binding sites in balanced blood will be occupied by O_2 and half by CO. This pollutant thus reduces the oxygen supply to the tissues, causing an insufficient blood supply to the heart and even circulatory diseases.

NO_2

The health effects associated with nitrogen dioxide are difficulty in breathing and greater respiratory infection. Short-term exposures to high NO_2 concentrations appear to be more harmful to human health than long-term exposures to moderate NO_2 concentrations.

HC

The significant effects of hydrocarbon pollution are eye irritation, visibility reduction, and plant damage. Although not all hydrocarbon pollution is immediately dangerous to human health, it is known that hydrocarbons and NO_x react under sunlight to produce ozone, which is very hazardous to the human body. Therefore, the control of hydrocarbon emissions is very important.

The conclusion reached by John Goldsmith (see John Goldsmith, M.D., in *Basics of R&D*, "The Health Effects of Air Pollution," published by the American Thoracic Society) is that the effects of individual pollutants are less damaging than those occurring from the interaction of several pollutants (e.g., SO_2 and particulates forming acid sulfate aerosols, and hydrocarbons reacting with NO_x

to produce ozone). In other words, the community should control all pollutants equally and not consider one pollutant to be more or less important than another. In accordance with this conclusion, state and federal air quality standards were established (Appendix A) and local agencies were made responsible for adopting and enforcing the rules and regulations necessary to protect public health. Industries must develop and select proper control equipment to comply with these rules and regulations.

Among many other technologies, the most common means of control are the following: process modification and material change, combustion, absorption, adsorption, condensation, scrubbers, and odor control. Each control technique is discussed briefly below. A final section summarizes criteria gaseous pollutant emission control.

CONTROL BY PROCESS MODIFICATION AND MATERIAL CHANGE

The air pollution of an industrial process can be reduced by modifying the process or changing process materials without sacrificing the original production capacity. Some examples are discussed below for this control method.

Process Modification

Process modification includes process change, process control, and equipment modification.

Process Change
Before the 1950s, coke was manufactured primarily by beehive coke ovens. This method pollutes by emitting organic vapors to the atmosphere as a result of the distillation involved in coke manufacturing. By the 1960s most of the beehive coke ovens in the United States had been replaced with by-product coke ovens (slot-type coke ovens). The by-product coke ovens normally are equipped with a chemical recovery system to collect organic vapors, which can be condensed to become coal tar (a black, thick, opaque liquid). Many synthetic compounds have been developed from this product, including medicines, explosives, and perfumes. These coal tar products are sold as by-products from the manufacture of coke (which explains the name of the ovens). The hydrocarbon emissions were reduced by this process change.

Process Control
Instead of process change, improved process control may be used to reduce or avoid air pollution. The rate of material feed, the uniformity of mixing, the

reaction temperature and pressure, and other factors influencing the reaction rate in a chemical process must be controlled at an optimum operating condition to avoid accidental periods of excessive discharge of contaminants. In some cases, H_2S is released to the atmosphere from chemical plants because of the failure of controls designed to keep the reaction rate within proper limits. Proper process control is an important means of reducing gaseous emissions.

Equipment Modification

Equipment modification belongs to the category of process modification. It includes changing designs, revising old models, and adding new auxiliary portions to equipment for certain purposes. One example is the modification of automobile engines to meet California's emission standards for hydrocarbons and carbon monoxide in exhaust gas. General Motors (GM) and Ford have designed vehicles on the principle of forcing air into the exhaust manifolds to complete combustion, which is another example of reducing emission by equipment modification.

Changing Process Materials

In some cases, process materials with high pollutant emissibility can be replaced by lower pollutant emissibility materials without downgrading production, thus reducing gaseous emissions. For example, in certain metal-cleaning operations, or in metal-coating industries, organic solvents may be replaced with less volatile agents, or solvent-based coatings replaced with water-based coatings. As a result, less volatile and less toxic solvents are emitted to the atmosphere.

CONTROL BY COMBUSTION

Many organic compounds released by manufacturers can be converted to harmless CO_2 and water by rapid oxidation (combustion). This can be achieved only when complete combustion takes place. Incomplete reactions may result in the formation of aldehydes (a special class of organic compounds that are highly toxic and strongly irritating to tissues), organic acids (such as acetic acid, which is highly toxic when ingested), carbon, and carbon monoxides. Factors affecting the completeness of combustion are temperature, oxygen, time, and turbulence, discussed below.

Temperature

Every combustible substance has a minimum ignition temperature that must be attained, or exceeded, in the presence of oxygen if combustion is to be assured under the given conditions. The ignition temperature is defined as the temperature at which more heat is generated by a reaction than is lost to the surroundings.

The ignition temperature of gases volatilized from coal is higher than the ignition temperature of the fixed carbon in coal. Before the ignition temperature of the fixed carbon is attained, the gaseous compounds in the coal are distilled off but not ignited. Therefore, if complete combustion of those gases is to be achieved, it is necessary that the temperature of the flue gases be raised to the ignition temperature of the gases. To achieve a sufficiently high temperature, it may be necessary to add auxiliary heat (through a gas-fired burner) to the flue gases. Because the reaction rate increases with temperature, a temperature above the ignition temperature of the combustible gases may be necessary to accomplish complete combustion in a reasonable amount of time.

Oxygen

The supply of oxygen determines the end products of combustion. For example, when methane (CH_4) is burned with too little oxygen, solid carbon (C) is produced. The solid carbon becomes soot and smoke. On the other hand, when methane is burned with a sufficient oxygen supply, the reaction is complete combustion, and there is no smoke.

Another example is the combustion of carbon. When carbon is burned with insufficient oxygen, carbon monoxide (CO) is produced. When carbon is supplied with sufficient oxygen, the end product of combustion will be carbon dioxide (CO_2).

To achieve complete combustion of a combustible compound, a stoichiometric quantity of oxygen—that is, a theoretically calculated amount of air—should be supplied. Practically, it is necessary to use more than the theoretical air required to assure sufficient oxygen for complete combustion; but the excess of air should be held to a practical minimum to avoid excessive heat loss through the stack.

Time

Time is an important function in the design of combustion equipment. The residence time and the time of combustion of particles both play major roles. (The residence time is the time that a particle resides in the equipment at combustion conditions, and the combustion time of a particle is the time required for combustion of the particle.) In order to achieve complete combustion, the residence time should be greater than the combustion time of the particle. The time of residence depends on the size of the equipment; the time of combustion is controlled by the temperature and aerodynamic factors (pressure, velocity, laminar flow, or turbulence). The smaller the unit is, the higher the temperature must be to oxidize the material at the time of contact between the particle and the flames.

Turbulence

The oxygen that is supplied should be mixed (to form turbulence) with the gas being burned so that the oxygen is available to the combustion substance at all times. Turbulent gas flow can be obtained either by nozzle mixing or by a mixing plate.

The effects of turbulence on particulate and gaseous combustions are different. If a substance burns without turbulence, the initial products of combustion act as a screen for the incoming oxygen, and thus slow the rate of surface reaction. The burning of gases requires thorough mixing of the gases with air; otherwise, separate zones will be formed between the gases and the air, and the gases and air will escape unburned or incompletely burned.

CONTROL BY ABSORPTION

In gas absorption a soluble component of a gas mixture is dissolved in a liquid. Water frequently has been used as a solvent to dissolve solutes such as CO, CO_2, NO_2, NH_3, H_2S, HCl (hyrdrogen chloride), HF (hydrogen fluoride), and HCN (hydrogen cyanide). There are other solvents for gas absorption, such as propylene carbonate, butoxy diethylene glycol acetate, and methoxy triethylene glycol acetate.

The device used for continuously contacting a liquid with a gas stream may be a tower filled with irregular solid packing material or an empty tower. The tower may contain a number of bubble-cup or sieve plates. The liquid is sprayed into the tower, and the gas streams flow countercurrently through the equipment in order to obtain the greatest rate of absorption.

Absorption is closely related to solubility and is governed by Henry's law. Solubility is the ability or tendency of one substance to mix uniformly with another. It can be a solid in a liquid, a liquid in a liquid, or a gas in a liquid, which is of concern here.

The physical chemistry of solubility is a mathematically complex subject involving Henry's law among other principles.

Henry's Law

Henry's law states that when a liquid and a gas remain in contact, the weight of the gas that dissolves in a given quantity of liquid is proportional to the pressure of the gas above the liquid. This is true only for equilibrium conditions (that is, the quantity of gas dissolved is no longer increasing or decreasing when enough time has passed, and when the partial pressure of the solute gas does not exceed 1 atmosphere). When Henry's law is true, the solubility is defined by using Henry's constant, H, and the temperature, where:

$$H = Pa/X_A$$

where:

Pa = partial pressure of the solute gas, A

X_A = mol fraction of the solute in solution

Henry's constant can be used to determine the amount of solute dissolved in a given amount of liquid through a known partial pressure, or vice versa.

Gas Absorption

Henry's law is involved in determininig the solubility of various gases in water at equilibrium, whereas in gas absorption the principle of diffusion plays an important role.

Generally, diffusion is the movement of materials between two phases caused by a vapor pressure difference or a concentration difference for materials in the two phases. It is characterized by the fact that material is transferred from one phase to another or between both phases. This phenomenon is called mass transfer or material transfer.

There are several types of diffusion: (1) homogeneous diffusion, (2) steady-state equimolal counterdiffusion, (3) steady-state diffusion of one component through a second stagnant component, and (4) unsteady-state diffusion. Homogeneous diffusion is the diffusion of a material, either gas, liquid, or solid, that contains two or more identical components whose concentrations vary from point to point so that transfer of mass will take place, causing the concentrations to become uniform. Steady-state equimolal counterdiffusion is diffusion such as the mixing of two gases in a fixed amount of space. Steady-state diffusion of one component through a second stagnant component is the absorption of a slightly soluble gas into a nonvolatile liquid. Unsteady-state diffusion is diffusion with a varying concentration gradient.

Packed Towers

There are two major types of gas absorption systems, one consisting of packed towers and the other of plate towers. A pack tower is filled with one of many available packing materials. A plate tower has a number of plates or trays arranged in such a way that gas is dispersed (or forced to pass) through a layer of liquid on each plate.

When the liquid or solvent is introduced at the top of the packed tower, the surface of the packing material is wetted by the solvent, forming a large area of liquid film that can come into contact with the solute gas. Normally, a packed tower has a countercurrent flow so that gas is introduced at the bottom and passes upward through the packing material. This gives the highest control efficiency because the solute concentration in the gas decreases when it rises through the tower, and there is always fresh solvent available for the absorbing reaction.

Plate Towers

A packed tower holds less water than a plate tower, has a low pressure drop, and is cheaper and easier to use when handling corrosive substances, as long as the diameter of the tower is less than 2 feet. However, because of the packed material, the packed tower is relatively heavy, and thermoexpansion or contraction may cause it to be crushed. It also is necessary for solid material to be removed from the packed tower; so it may not be preferred. For large units, where solid material needs to be removed, or when thermoexpansion of the shell may occur, a plate tower should be used.

Generally, plate towers consist of a shell and plates or trays. The liquid flows in from the upper section of the tower and passes through the plates stepwise down to the bottom of the tower. Plate towers are chosen for large-scale operations (because they are cheaper), for low liquid rate operations, and where internal cooling is required; packed towers usually are selected for corrosive materials, a low pressure drop, small-scale operations, and liquids that foam badly.

Recently, a mist or fog tower has become commercially available. Solvent is sprayed through nozzles by means of compressed air producing an ultrasonic spray with droplets in the range of 10 microns. These droplets are suspended inside the tower, and when the contaminated gases are guided concurrently with the spray, the gas–liquid contact time can be extended to 30 or more seconds. If the solvent is a chemical selected to react with a particular contaminant, the efficiency will increase tremendously. The amount of spent solvent is so small that it can be oxidized, adsorbed, or biologically degraded.

CONTROL BY ADSORPTION

The previous section discussed absorption control, a means of dealing with liquid–gas interactions. This section discusses solid–gas contact processes known as adsorption control. Contaminants removed by adsorption are concentrated on a solid phase, whereas those separated by absorption are distributed in the bulk of an interacting liquid.

Definition of Adsorption

Adsorption is the capture and retention of a component from the gas phase by the internal and external surfaces of the adsorbing solid. The component captured is called the adsorbate, and the adsorbing solid is the adsorbent.

General Phenomena of Adsorption

Adsorption occurs in several overlapping steps: the first step is diffusion of the contaminant from the bulk gas through the boundary layer to the internal surface of the solid; the second step is the migration of the adsorbate from the external surface of the adsorbent to the surface of the pores; and the third step is the adsorption of particles onto the active side of the pores.

Several technical terms describe the adsorption process:

- *Residence Time:* This is a finite time when a molecule is retained on a solid surface. During this time, there is an exchange of thermal energy between the molecule and the surface.
- *Adsorption force:* Two types of forces can cause adsorption: intermolecular forces (van der Waals or molecular interaction forces), causing physical adsorption; and chemical forces (which usually involve electron transfer between the solid and the gas), causing chemical adsorption.
- *Adsorption rate:* The adsorption rate indicates the degree of surface coverage of the adsorbate on the adsorbent. It is determined by one or more diffusion factors:

  1. Diffusion from the bulk flow to the external surface of the adsorbent.
  2. Reaction of the phase boundary.
  3. Diffusion in the adsorbed surface layer.
  4. Pore diffusion.
  5. The residence time of these molecules on the surface.

- *Types of Adsorption:* There are two types of solid–gas phase interactions. In one there is random mixing of the two phases (static adsorption); in the other the motion of each phase is directed toward that of the other (dynamic adsorption).

Static Adsorption

Although most applications of air pollution control are those of dynamic adsorption, where adsorption equilibrium is not reached, it is important to know the equilibrium (static adsorption) conditions; modified equilibrium effects will influence a dynamic (nonequilibrium) system directly with respect to the number of molecules leaving the surface and going into the gas phase. When the gas molecules stay on the solid surface (adsorbent), there is an exchange of energy between the phases. If the time of adsorption is long enough, there will be thermal equilibrium; no net energy change will exist between them. The temperature then will remain constant, in an adsorption process known as isothermal adsorption.

Dynamic Adsorption

In general air pollution control processes, the two phases (gas–solid) often are directed toward each other; that is, dynamic adsorption systems often are used. With few exceptions, most adsorption systems have a greater heat of adsorption than the heat of evaporation of the same substance; that is, the entropy of the gas molecules of contaminated air adsorbed on the solid surface of a fixed bed (adsorbent bed) is greater than the entropy of the molecules in their original (liquid or solid) state. Greater entropy means a greater degree of freedom; so the adsorption process generally is exothermic.

When the contaminated air first passes through the bed, most of the contaminants (or adsorbates) are adsorbed at the inlet part of the bed, and the air passes on. When the inlet end of the adsorber becomes saturated, adsorption takes place farther along the bed. The bed length covering the range from the initial deposit to saturation of the adsorbate is called the mass transfer zone (MTZ) or the material transfer zone (MTZ). When more contaminated gas is passed through the bed, the mass transfer zone moves forward to a certain point, at which the exit concentration of the contaminant begins to rise rapidly above a certain given limit. This point (in time) is known as the breakthrough point, and the bed is fully saturated. The exit concentration is practically the same as that of the inlet.

The capability of dynamic adsorption is measured by the breakthrough capacity of the material used, which depends on the operating temperature, inlet concentration, flow rate, and bed depth. The length of the bed should be equal to or greater than the MTZ.

Adsorption Systems Applied to Air Pollution Control

Single-Unit Adsorption Systems

The simplest adsorber has a single bed in an enclosure, a horizontal or a vertical vessel. The vertical adsorption arrangement is used to treat a small amount of gas, whereas the horizontal arrangement has a large gas capacity. For greater efficiency, two or more beds are installed in the single-unit adsorber (once the adsorbate is saturated on the adsorbent, a new adsorbent should be used). The single-unit adsorber is used in air-conditioning systems and in the treatment of low solvent concentrations.

Multiple-Unit Adsorption Systems

Continuous operation requires at least two adsorber units. The gas-contaminated air enters the first adsorber and passes downward through the adsorbent (carbon bed), where the gas vapor is adsorbed. The vapor-free air passes out to the atmosphere, while a second adsorber strips the vapor of its adsorbents. Usually hot steam is passed through the carbon bed, heating the carbon to a temperature at which the vapor is desorbed and released.

This process is called regeneration of adsorbents. The stripping cycle must be long enough to allow the adsorbent to cool before it is returned to the adsorption process. Three units usually are required: an adsorbing unit, a cooling unit, and a regenerating unit. This multiple-unit adsorption system has been used for handling high concentrations of solvents, gasoline, and fuel oils.

Adsorbents Applied to Adsorption Systems

Activated charcoal (or carbon) is the most universally used adsorbent in adsorption systems. The activated carbon is made by destructive distillation of wood or peat, followed by heating the product to a high temperature with steam. It is used to collect, catch, and concentrate the adsorbate, usually a nonpolar compound, such as a hydrocarbon. Other adsorbents, such as activated silica and activated alumina, are used for polar compounds (water, sulfur dioxide, or alcohol). Activated carbon can be used to adsorb acidic gases (such as SO_2, HCl, or NO_2) by being impregnated with a suitable alkaline material. Impregnated materials enable more chemical reactions to be utilized.

Activated carbon comes in many forms used for various purposes, such as liquid purification, decolorization, and air purification. The selection of activated carbon depends on the type of contaminant, the concentration of the gas, operating conditions, the removal efficiency required, and the contact time.

Silicates of aluminum and sodium or calcium are known as zeolites, also called molecular sieves. A zeolite has a crystalline structure of approximately 50 percent internal porosity, with almost uniform pore diameters. Zeolites have been used to separate petroleum products in refining operations and as adsorbents to capture gaseous pollutants, depending on the size and shape of the gas molecules.

CONTROL BY CONDENSERS

A condenser converts vapors into liquid form. When air contaminants are emitted into the atmosphere as vapors, condensers can serve as control equipment. When heat is removed from a vapor, condensation occurs, and the resulting liquid may be disposed of in deep wells or treated separately. Additionally, condensers can be used as elements of integrated control systems to reduce the load on more expensive control devices, and they can remove vapor components that may affect the operation of major control devices or cause corrosion problems.

Types of Condensers

Condensers can be classified according to different principles of heat removal as surface condensers and contact condensers. In surface condensers, the moist vapor of the condensate and the coolants are separated by tube surfaces and are not in contact with each other. There are three types of surface condensers: tube-and shell condensers, air-cooled condensers, and atmospheric (natural draft) condensers. Contact condensers remove heat by using coolants (water) in direct contact with the condensing vapors. There are three types of contact condensers: simple spray chamber condensers, single high velocity jet (venturi type) condensers, and multiple high velocity jet condensers.

Surface condensers offer the advantages over contact condensers that less water is required, less condensate is produced, and the condensate is reusable. On the other hand, they need more auxiliary devices, thus necessitating more maintenance. Contact condensers are simple, more flexible, and less expensive to install than surface condensers, but their condensate cannot be reused, and waste disposal problems occur with their use.

Because their operating cost (water consumption and waste disposal) is high and their application is limited, development efforts for contact condensers have been limited in recent years. Surface condensers are better developed, but most of them use a film type of condensation because the film's heat transfer coefficient is only one-tenth that of the droplet. The sizing and the design of surface condensers have been well documented in other literature; so no attempt is made to describe them here.

CONTROL BY SCRUBBERS

Certain gaseous emissions are controlled by scrubbers. The gaseous emissions from industrial sources include HC1 from steel plants and textile facilities; HF and NO_x from metal finishing and aluminum plants; Cl_2, ClO_2, and CSO_2 from pulp and paper mills; and H_2S and CS_2 from food processing and textile facilities. These emissions are controlled by scrubbers using impingement, nucleation, absorption, or combination methods.

As described in Appendix A, in the impingement method targets are placed in the path of an air stream so that particles impinge upon the targets; the contaminants are intercepted and then washed away. In nucleation submicron particles are moistened to lower their temperatures; the water on the particles condenses, and the particle size increases so that the particles can be removed by impingement. As described earlier, absorption is the transfer of a gaseous emission from an exhaust mixture into a liquid. The main factors affecting the absorption process are the solubility of the gas being removed in the scrubbing liquid and the means of obtaining contact between the gas and the liquid. Usually, water is used as the scrubbing liquid for hydrogen fluoride (HF) and hydrogen chloride (HC1). A sodium hydroxide (NaOH) scrubbing liquid is reacted with chlorine gas (Cl_2) to produce sodium hypochlorite (NaOCl), which is soluble in H_2O.

For a certain type of scrubbing liquid (H_2O) and a constant concentration of the gaseous contaminant, the collecting efficiency of the scrubber is a function of the liquid flow rate (gpm). The efficiency also depends upon the type of scrubber or the means by which contact between the liquid and the gases is promoted.

Cross-Flow Scrubbers

The most common scrubber used for gaseous emission control is the cross-flow scrubber (named for the flow of water and air across each other). The air stream moves horizontally through a packed bed; the scrubbing water flows vertically through the packing. It is designed to remove particles and gaseous pollutants.

A cross-flow packed scrubber may be modified as follows:

- *Parallel-flow scrubber:* A front spray prevents a solid buildup on the packing, thus reducing the pressure drop. Water spray nozzles are installed upstream of the packed bed as front washer sprays, so that the gas stream and the scrubbing liquid reach a parallel flow.
- *Deep-bed scrubber:* This is a modified two-stage horizontal scrubber. It involves adding more packing, placing pipes at the top of the packing for water distribution, and removing the rear nozzle, to obtain better water distribution and better or improved performance.

Vertical (Packed) Scrubbers

Gas flow in a vertical-type scrubber is in the vertical direction. The water distributor is installed at the upper portion of the packing, producing a mist-elimination effect at the packing above the water distributor and a deep-bed effect at the packing below the water distributor.

Vertical packed scrubbers may be modified as follows:

- *Countercurrent packed scrubber:* Here the liquid separator (eliminator) and the packing are separated. Gas flows upward and the scrubbing liquid downward.
- *Two-stage plate tower scrubber:* In the two-stage plate tower scrubber, the packing is separated into two sections (Stage 1 and Stage 2), and the mist eliminator is at the top.

ODOR CONTROL

When gases and vapors exceed a certain concentration range, they usually have characteristic odors, unless they are the original components of air (oxygen, nitrogen, carbon dioxide, water vapor, and argon).

Gases and vapors causing major odor problems are from the following sources:

1. Animal odors
   - Meat packing and rendering plants
   - Fish-oil factories
   - Poultry processing
2. Combustion odors
   - Gasoline- and diesel-engine exhaust
   - Coke-oven and coal-gas emissions
   - Improperly adjusted heating systems
3. Food processing odors
   - Coffee roasting
   - Restaurant cooking
   - Bakeries
4. Paint-related industrial odors
   - Paint, lacquer, and varnish manufacturing
   - Paint spraying
   - Solvent usage
5. General chemical odors
   - Hydrogen sulfide (H_2S)
   - Sulfur dioxide (SO_2)
   - Ammonia (NH_3)
6. General industrial odors
   - Burning rubber
   - Dry-cleaning
   - Fertilizer
   - Asphalt roofing and street paving
   - Asphalt manufacturing
   - Plastic manufacturing
7. Foundry odors
   - Coke-oven operation
   - Heat treating
   - Smelting
8. Combustible waste odors
   - Backyard trash fires
   - City incinerators burning garbage
   - Open-dump fires

9. Refinery odors
   - Crude oil processing
   - Gasoline production
   - Sulfur recovery
10. Decomposition waste odors
   - Meat oxidation
   - Protein decomposition
   - Plant-cell decomposition
11. Sewage odors
   - Sewage systems
   - Sewage treatment plants

Some gases and vapors have pleasant odors; others are irritants and have unpleasant odors. Communities consider the latter to be objectionable atmospheric contaminants.

The quality of an odor can be characterized as: fragrant, putrid, musky, phenolic, pleasant, foul-smelling, penetrating, or poisonous. Odor quality has been predicted by using a reference standard, known as the detection threshold, indicating the minimum concentration at which an odorous substance can be distinguished from odor-free air. The threshold depends upon the substance and the recipients' sensitivity to its odor. A cubic foot of air at the threshold is defined as "one odor unit"; a cubic foot of a high-level odorant is said to contain "n odor units," equivalent to diluting the odorant to n cubic feet at the threshold level. The "50% threshold" is the concentration at which the odor can be recognized by 50 percent of the population.

The strategies used to reduce the odor level can be categorized as odor aversion and odor control.

Common Odor Aversion

The methods used to avert common odor include the following: ventilation, dispersion, reodorization, process conversion, and source modification.

Ventilation
Odor ventilation is the most common method of removing odorous air from enclosed spaces. Ventilation rates are as follows:

- To maintan the regular air levels: 4 ft^3/min/person.
- To ventilate odorous sources: 40 ft^3/min/person (tobacco, food, cosmetics, people).

Ventilation cannot be used when the quality of the outdoor air is unsatisfactory.

Note also that the odor depends on the input and the output rate of the ventilated air, but is independent of the volume of air affected.

Dispersion

An odorous gas can be dispersed through a stack and diluted into the atmosphere. In the equation for calculating its dispersion (not included here), the ground-level concentration is a function of the concentration of the odorant in the emitted gas stream and the average velocity of the atmospheric air flow (wind velocity and atmospheric turbulence). The calculation yields only an average value. In order to avoid any momentary exposure to an unacceptable odorous gas, the degree of odor reduction should be greater than the calculated odor emission.

Reodorization

Reodorization adjusts odors by mixing them with more pleasant odorants without causing chemical changes, such as the use of perfumes for body odors, tobacco smoke, and so on.

Process Conversion

Process conversion changes an odor-producing process to an odor-controlled process. There are several approaches, involving temperature, pressure, volume, maintenance, and housekeeping.

Temperature
Excessive temperatures during the drying of a material may produce odorous decomposition products. Insufficient furnace temperatures (in the burning of odorous waste gases) may produce products that are more odorous than the original material.

Pressure
Altering the environment of a process (such as going from a slightly positive pressure to a slightly negative pressure) by changing a damper position or moving a fan location may reduce the leakage of odorous material.

Volume
Increasing the ventilation volume will tend to dilute odorous emissions; but if the odorant is a vaporized liquid, increasing the ventilation volume will increase the quantity of odorant emitted. Decreasing the ventilation volume will reduce odor levels downstream; however, volume reduction can be achieved only when explosion hazards have been adequately considered.

Maintenance
If odorants are released from the leakage of flange seals or uncovered tanks not in use, good maintenance practices can prevent the odors.

Housekeeping
Accumulations of volatile wastes and food by-products are sources of odorous emissions. Good housekeeping can eliminate them.

Source Modification
The goal of this control method is to change odorous sources to less odorous ones, which can be done by adding odor modifiers. Alcohols, aldehydes and esters have been used to modify processes, for example, at meat-scrap and bone-rendering plants, so that the odor concentration ranges from 1 ppm to 10 ppm.

If possible, the odor-polluting equipment or source should be located far from residential areas.

Specific Odor Control

The following equipment may be added to processes to control odor: scrubbers, combustors, absorbers, adsorbers, and condensers.

Scrubbers
Scrubbers have been used to remove odors from air (air scrubbers). One of several mechanisms may be involved.

- The odorous vapor may dissolve in a scrubbing liquid.
- The odorous vapor may condense in a cold liquid.
- Particulate odorants may be scrubbed from the air.

The scrubbing liquid may be water, a solution of chemical reactants (oxidizing agents) in water, or a glycol solution (alcohol). The oxidizing agents may be chlorine (Cl_2), chlorine dioxide (ClO_2), potassium permanganate ($KMnO_4$), or sodium hydroxide, caustic soda ($NaOH$). The solution choice depends on the offending odor.

Combustor
A combustor burns odorous gases by one of the following methods: flame oxidation, catalytic oxidation, or chemical oxidation.

Flame Oxidation
Odorous air is mixed with combustion gases and heated to 1100 to 1500°F in an

incinerator. The final products are H_2O and CO_2 (odorless). Partial combustion may increase the odor. The flame retention time is usually $\geq$ 0.3 second.

Catalytic Oxidation

Odorous air is passed through catalytic units at 500 to 800°F. The method's effectiveness may be reduced because of catalyst poisoning (by lead), deposition of carbon due to incomplete combustion, or loss of catalyst due to abrasion by air-borne particles.

Chemical Oxidation

The odorant is mixed with oxidizing agents such as ozone, permanganates, chlorine, and chlorine dioxide. Ozone is used to deodorize exhaust gas in stacks. Potassium permanganate is used to deodorize sulfur compounds. Chlorine (Cl_2), chlorine dioxide (ClO_2), or chlorine plus H_2O controls fishmeal odors.

Absorbers

Many odors are removed by passing the odorous gases through certain liquids or solutions. Examples are H_2S (hydrogen sulfide), HCl (hydrogen chloride), and NH_3 (ammonia). Hydrogen sulfide has an offensive odor, and is a by-product of petroleum refining. Hydrogen chloride has a suffocating odor, and is produced by metallurgical plants. Ammonia has a sharp, intensively irritating odor, and is a product of the zinc-galvanizing process (a process of coating a thin layer of zinc on clean steel by immersion in molten zinc, for which zinc ammonium chloride solution is used, producing NH_3 and HCl). Although H_2S is soluble, and both HCl and NH_3 are very soluble in water, none of these is removed from the mixture gas by contacting water because that method is not effective enough to reduce odors.

Both H_2S and HCl are removed by amine solutions. An amine is a derivative of ammonia (NH_3) in which one of the hydrogens has been replaced by an organic radical containing H and C (for example, CH_3) atoms (CH_3NH_2 is an amine solution). All amine solutions are basic or alkaline in nature, with their pH in the 7 to 14 range. They combine readily with H_2S, HCl, and other strong acids.

As has been discussed, amine absorption (or Girbotol absorption) removes H_2S from a gaseous mixture. An organic amine (for example, ethanol amine, $HOCH_2CH_2NH_2$, which is basic) flows down a path through a tower where it is in contact with gaseous mixtures and absorbs (acidic) H_2S. The amine, contaminated by H_2S, is sent to a steam stripper, where it flows countercurrently to the steam. Then the amine is returned to the tower for reuse. Both H_2S and sulfur can be recovered, and in the case of HCl removal both HCl and chlorine can be recovered.

Ammonia can be absorbed by a mixture of water and caustic (alkali) soda (such as sodium hydroxide).

Adsorbers

Adsorbers are used mainly to remove odorants from organic gases, such as food waste gases. Activated carbons, in granular form in a fixed bed of cylindrical adsorbers, are the most common adsorbent. When a carbon bed becomes saturated, it can be regenerated by steam stripping. The collected odorous vapor, together with exhaust steam, is passed through a condenser, where the organic fraction is separated from the steam. After the reactivation stage, the hot carbon bed should be cooled before being returned to the adsorption process because carbon is a catalyst for the decomposition of organic matter, which may result in the formation of nuisance odors.

Condensers

Condensers are particularly useful for odor control when the effluent contains a large portion of water vapor, as in rendering plants. Condensation of the water may reduce the concentration of odorants. As noted previously, there are two types of condensers: surface condensers and contact condensers. Contact condensers are widely used for odor control applications because they are less expensive and have a higher degree of control than surface condensers. However, their use has also led to problems concerning contaminated water and the re-release of odors.

SUMMARY OF CRITERIA GASEOUS POLLUTANT EMISSION CONTROL

Criteria gaseous pollutants include ozone (its precursors are hydrocarbons and NO_x), oxides of sulfur, oxides of nitrogen, and carbon monoxide. The major emission sources and the common control methods for the individual pollutants are summarized below.

Hydrocarbons

The major sources of hydrocarbon emissions are the petroleum refining industry, gasoline service stations, process coating lines, dry cleaners, degreasers, and the chemical industry. The control technologies include the following:

1. Process and materials changes
   - New degreasing solvents and techniques
   - Water-base coatings
   - Powder coatings
2. Add-on control devices
   - Incinerators
   - Absorbers

- Adsorbers
- Condensers
- Catalytic oxidizers
3. Cover-seals
- Floating roofs for petrochemical tanks

Oxides of Sulfur

The major source contributors are the petroleum refining industry, electric utility plants, and steel industries. The methods of controlling oxides of sulfur (SO_x) emissions include wet scrubbing, dry caustic injection, dry adsorption, and dry catalytic methods, which are discussed briefly below.

- In a wet scrubbing method, lime (CaO) or limestone ($CaCO_3$) slurries or another alkaline solution is used to react with SO_x, forming calcium sulfate ($CaCO_3$), which is gypsum and can be used as plaster.
- A dry caustic injection method injects limestone powder into flue gas. The resulting sulfate material ($CaSO_4$) can be collected in a baghouse.
- An adsorption method uses charcoal-activated carbon in granular form, which is set (in contrast to a fixed bed) in a continuous bed moving from one chamber to another in a closed loop. When the contaminated gases enter the first (reaction) chamber, sulfur oxides adhere to the charcoal, which is moved to the second chamber for regeneration. Steam is sprayed on the sulfur-coated charcoal to release the sulfur oxides, and the regenerated charcoal is returned to the reaction chamber for reuse. The freed sulfur oxides are ducted to the sulfur recovery facility to reclaim the sulfur element. This is a continuous process; no interruption occurs between the reaction and regeneration steps.
- A dry catalytic method uses a metallic catalyst, such as copper oxide coated on an aluminum substrate, to contact SO_x-laden gases. Copper oxide reacts with SO_x and oxygen, forming a copper sulfate product on the fixed substrate. During the regeneration of the catalyst, hydrogen gas is used to react with copper sulfate, yielding sulfur dioxide (SO_2) and water vapor. Steam is used to concentrate elemental sulfur at the sulfur recovery unit. The regenerated copper oxide catalyst can be sent back to the process for reuse.

Oxides of Nitrogen

The major sources of oxides of nitrogen (NO_x) emissions are electrical utility plants, petroleum refineries, chemical plants, steel industries, and internal combustion engines. To control NO_2 emissions, the industries use low-NO_x burners, thermo-de-NO_x, or selective catalytic reduction methods.

- Low-NO_x burners are either fuel or air staged burners. Their primary purposes are to limit the adiabatic peak temperature, to complete combustion, to dilute the combustion product, and to reduce NO_x emissions.
- A thermo-de-NO_x method utilizes a spray of ammonia or urea directed into the flue gas. In the presence of hydrogen, the flue gas temperature will need to be approximately 1500°F. Within a very narrow band (approximately ±100°F) around the designated temperature, nitrogen oxides and ammonia react with oxygen to form molecular nitrogen and water vapor. If the flue gas temperature exceeds the upper range, ammonia reacts with oxygen and yields nitrogen oxides. If the temperature drops below the lower temperature range, the ammonia reaction rate decreases to a minimum. Keeping the flue gas temperature within the specified range throughout the process is cumbersome and impractical.
- Using a selective catalytic reduction (SCR) method to control NO_x has proved to be the most effective approach to NO_x reduction. The catalyst can be copper sulfate, titanium dioxide, or other compounds (such as vanadium). The catalyst pellets are set on a moving bed, which is circulated between the reactor and the catalyst generator. (For a fixed bed, a zeolite-based catalyst may used for maximum control efficiency). Ammonia is sprayed into the flue gas before it enters the reactor. A nitrogen oxides—ammonia reaction is promoted by the catalyst, producing molecular nitrogen and water vapor. When the catalyst pellets reach the regenerator, hydrogen gas frequently is used to regenerate them for reuse. The SCR reaction can take place at flue gas temperatures from as low as 300°F up to 900°F, a much wider range than that of the thermo-de-NO_x method.

Carbon Monoxide

The major sources of carbon monoxide are any incomplete combustion reaction, chemical reaction by-products, and coal gasification. The common control method uses a precious metal, such as platinum, as a catalyst to oxidize carbon monoxide to carbon dioxide, or supplies sufficient oxygen to complete the combustion reactions.

Select Bibliography

Air Toxics "Hot Spots" Program, Risk Assessment Guidelines, 1991. California Air
Pollution Control Officer Association.

American Gas Association Research Report, 1990. "Natural Gas as an Alternative
Transportation Fuel: Clean Air Strategy for State and Local Officials."

Appleby, A. J., et al. *Fuel Cell Handbook.* New York: Van Nostrand Reinhold, 1989.

ASHRAE Guide and Data Book, 1967. "Handbook of Fundamentals."

ASHRAE Guide and Data Book, 1967. "Systems and Equipment."

Assembly Bill 2595, State of California, 1988. "California Clean Air Act."

Becker, E. R., et al. "Catalyst Design for Emission Control of Carbon Oxide and
Hydrocarbons from Gas Engines." *The 81st Annual Conference of APCA,* 1988.

Bretherick, L. *Handbook of Reactive Chemical Hazards.* London: Butterworths, 1977.

Brunauer, Stephen. *The Adsorption of Gases and Vapors.* Princeton, NJ: Princeton
University Press, 1943.

Bundy, McGeorge. *Nuclear Power Issues and Choices.* Boston, MA: Ballinger Publishing, 1977.

California Environmental Quality Act—1970. State of California, 1970.

Cheremisinoff, Paul N., et al. *Pollution Engineering Practice Handbook.* Ann Arbor,
MI: Ann Arbor Science Publishers, Inc., 1975.

Chiras, D. D. *Environmental Science.* Menlo Park, CA: The Benjamin/Cummings
Publishing, 1985.

Cichanowicz, J. E. "Selective Catalytic Reduction for Coal Fired Power Plants: Feasibility
and Economics." *EPRI Report CS-3603, RR 1256-7,* 1984.

Conference Record, 1985. "The Seventh Annual Science, Engineering, and Technology
Seminar." Houston, Texas.

Considine, Douglas. *Chemical and Process Technology Encyclopedia.* New York:
McGraw–Hill, 1974.

Crowser, K. E., et al. *Synthetic Fossil Fuel Technology.* Ann Arbor, MI: Ann Arbor
Science Publishers, 1980.

Danielson, John. *Air Pollution Engineering.* Research Triangle Park, NC: EPA, 1973.

Das Saarbruchen Zukunstskonzept Energie, 1992. Stadtwerke Saarbruchen, A. G.

DeLorenzi, Otto. *Combustion Engineering.* New York: Combustion Engineering Inc., 1957.

Encyclopedia of Chemical Technology. New York: John Wiley & Sons, Inc., 1992.

Encyclopedia of Science and Technology. New York: McGraw–Hill, 1982.

EPA Document AP42, 1981. "Compilation of Air Pollutant Emission Factors."

EPA Document 450/2-78-125a, 1979. "Stationary Internal Combustion Engines."

EPA Document 450/2-81-005. 1973. "Control of Gaseous Emissions."

EPA Document 450/3-82-006a, 1982. "Fossil Fuel Fired Industrial Boilers, Background Information."

EPA Document 450/4-92-008a, 1992. "User's Guide for Industrial Source Complex (ISC II), Gaussian Dispersion Models."

EPA Document 600/7-79-205, 1979. "NO$_x$ Abatement for Stationary Sources in Japan."

EPA Document 600/8-82-014, 1982. "PTPLU—A Single Source Gaussian Dispersion Algorithm."

EPA Document 625/6-91-014, 1991. "Handbook on Control Techniques for HAPs."

Evans, Ulick. An Introduction to Metallic Corrosion. London: Edward Arnold Ltd., 1981.

Federal Clean Air Act–1970 Amendment. Washington, D.C.: U.S. Government, 1970.

Federal Clean Air Act–1990 Amendment. Washington, D.C.: U.S. Government, 1990.

Felbeck, David K., and G. Anthony Atkins. *Strength and Fracture of Engineering Solids.* Englewood Cliffs, NJ: Prentice Hall, Inc., 1984.

Fuel Flue Gases. American Gas Association, Arlington, Virginia, 1941.

Fultz, Keith, et al. "Nuclear Waste—Quarterly Report on DOE's Nuclear Waste Program as of September 30, 1988." *GAO/RCED-89-22FS.* United States General Accounting Office; November, 1988.

Gary, James H., et al. *Petroleum Refining Technology and Economics.* New York: Marcel Dekker, Inc., 1984.

Gas Research Institute Research Report GRI 87/0265, 1987. "Environmental Benefits of CNG-Fueled Vehicles."

Gas Research Institute Research Report GRI 89/0249, 1989. "Assessment of Environmental, Health, and Safety Issues Related to the Use of Alternative Transportation Fuels."

Grayson, Martin. *Kirk—Othmer Encyclopedia of Chemical Technology.* New York: John Wiley & Sons, Inc., 1979.

Haagen-Smith, A. J. *Reactions in the Atmospheres.* New York: Academic Press, 1962.

Hanf, Edward. *Environmental Science and Technology,* 1970.

Hassler, John. *Active Carbon: Activated Carbon.* New York: Chemical Publishing Company, 1963.

Hawley's Condensed Chemical Dictionary, Eleventh Edition. New York: Van Nostrand Reinhold Company, 1990.

Hazen, Robert. *Breakthrough.* New York: Summit Books, 1988.

Healy, Timothy, et al. *Energy and Society.* San Francisco: Boyd & Fraser Publishing Co., 1983.

Hesketh, Howard. *Air Pollution Control*. Ann Arbor, MI: Ann Arbor Science Publishers, 1979.

Industrial Ventilation, 1986. Committee on Industrial Ventilation.

International Critical Tables. New York: McGraw–Hill, 1928.

Johnstone, H. F., and M. H. Roberts. "Deposition of Aerosol Particles from Moving Gas Streams." *Industrial Engineering Chemistry*. November, 1949.

Kinney, Patrick, et al. "The Health Effects of Methanol Vapors: The Health Effects Institute's Analysis." The 81st Annual Meeting of APCA, Proceedings. Air Pollution Control Association, 1988.

Kinoshita, K., et al. *Fuel Cell Handbook*. U.S. Department of Energy, 1989.

Kobe, Kenneth, et al. *Thermochemistry for the Petroleum Industry*. Houston: University of Texas Press, 1949.

Land, Herbert. *Industrial Pollution Control Handbook*. New York: McGraw–Hill, 1971.

Landolt-Boernstein, A. *Physikalische-Chemische Tabellen*. Berlin, Germany, 1927.

Langone, John. *Superconductivity: The New Alchemy*. Chicago: Contemporary Books, 1989.

Lee, Arthur S. "Expanding a New Horizon of Carbonaceous Resources for the Increasing Population." *(unpublished paper)* 1987.

Lee, Arthur S. "Formulation of Fossil Fuels." *(unpublished paper)* 1987.

Lee, Arthur S. "Health and Environmental Problems of Organic Contaminants in Shale Oil and Oil Shale." *(unpublished paper)* 1987.

Lee, Don. "Using Impregnated Activated Charcoal." *Journal of the American Association for Contamination Control*, 1965.

Leva, Max. *Tower Packings and Packed Tower Design*. Akron, OH: The United States Stoneware Co., 1953.

Levenspiel, Octave, et al. "A Numerical Solution to Dimensional Analysis." *Industrial and Engineering Chemistry*, February, 1955.

Makonsi, Jason. "Reducing NO_x Emissions." *Power*, September, 1988.

Manahan, Stanley E. *Environmental Chemistry*. Boston: Willard Grant Press, 1979.

Martin, A. E. *Emission Control Technology for Industrial Boilers*. Noyes Data Corporation, 1981.

McLean, Kenneth. *Experimental Results of Surface Resistivity vs. Water Vapor Pressure at Various Temperatures*. Wollongong, Australia: University of Wollongong Press, 1970.

Mikovich, J. J. "What's Happening to Pyrolysis?" *Pollution Engineering*, March–April, 1972.

Miller, Tyler G. *Living in the Environment*. Belmont, CA: Wadsworth Publishing, 1985.

Moran, Joseph, et al. *Introduction to Environmental Science*. San Francisco: W. H. Freeman & Co., 1980.

National Research Council Report. "Odors from Stationary and Mobile Sources." Washington, D.C.: National Academy of Science, 1979.

Nelson, W. L. *Petroleum Refinery Engineering*. New York: McGraw–Hill, 1969.

Oglesby, Sabert, Jr., et al. *A Manual of Electrostatic Precipitator Technology*. Birmingham, AL: Southern Research Institute, 1970.

Oglesby, Sabert, Jr., et al. *Electrostatic Precipitator*. New York: Richard A. Young, Publisher, 1978.

Perry, H. Robert, et al. *Chemical Engineers' Handbook*. New York: McGraw–Hill, 1973.

Pitts, James Jr., et al. "Photochemical Smog." *ISAP*, 1972.

Power. "Evaluation of Expansion-Joint Behavior." January, 1961.

Rafson, Harold. "On Quad's Mist Scrubbing." Seminar, Los Angeles, 1992.

Rich, Gerald. *Pollution Engineering*. Northbrook, CA: Pudvan Publishing, 1986.

Rosin, Rammler, and Intelmann. *Zeitschrift Verein Deutsche Ing*. Vol. 76, 433–437.

Rules and Regulations. South Coast Air Quality Management District, Diamond Bar, CA, 1992.

Seidell, Atherton. *Solubilities of Inorganic and Organic Compounds*. New York: Van Nostrand Reinhold, 1992.

Sherwood and Pigford. *Adsorption and Extraction*. New York: McGraw–Hill, 1952.

Shreve, R. N., et al. *Chemical Process Industrials*. New York: McGraw–Hill, 1977.

Shore, D. E., et al. "Haynes Unit 4, Thermal De NO_x Test Program, Oil Fuel." *KVB Report 71 33640-2060*. Los Angeles, Western Engineering Division, 1984.

Simon, Randy, et al. *The Path of No Resistance: The Story of the Revolution in Superconductivity*. New York: Plenum Press, 1989.

Singer, Joseph. *Combustion Fossil Power Systems*. Combustion Engineering Co., 1981.

Smith, Charles O. *The Science of Engineering Materials*. New York: Prentice Hall, 1977.

Steele, William. *The Physical Adsorption of Gases on Solids*. Amsterdam: Elsevier Publishing, 1967.

Symposium on Stationary Combustion Nitrogen Oxide Control. Radian Corporation, California, 1987.

The North American Combustion Handbook. The North American Manufacturing Company, 1953.

Timoshenko, Stephen. *Elements of Strength of Materials*. Princeton, NJ: Van Nostrand, 1968.

Treybal, Robert. *Mass Transfer Operation*. New York: McGraw–Hill, 1968.

Turk, Amos. "Industrial Odor Control and Its Problems." *Chemical Engineering*. November, 1969.

Turner, J. H., et al. "Sizing and Costing of Fabric Filters." *Journal of the APCA*. June, 1987.

Van Splinter, Gregg A. "Recovery and Reclamation of Nuclear Waste." *(unpublished paper)* 1981.

Vatavuk, W., et al. "Estimate the Size and Cost of Baghouse." *Chemical Engineering*. March, 1982.

Wang, Lawrence, et al. *Handbook of Environmental Engineering*. Totowa, NJ: The Humana Press, 1980.

Waneilista, Martin P., et al. *Engineering and the Environment*. Brooks/Cole, The Engineering Division, 1984.

Werne, Roger, et al. "Mirror Fusion." *Mechanical Engineering*. July, 1981.

White, Harry. *Industrial Electrostatic Precipitator*. Reading, MA: Addison–Wesley Publishing, 1963.

Williams, E. T., et al. *Stoichiometry for Chemical Engineers*. New York: McGraw–Hill, 1958.

Young, D. M., et al. *Physical Adsorption of Gases*. London: Butterworths, 1962.

Zeldovich, J. "The Oxidation of Nitrogen in Combustion and Explosions." *ACTA Physicochim*. Moscow, U.S.S.R., 1946.

Zeolite/Ceramic Molecular Sieve (SCR) NO_x Abatement System, Stueler, Cer-NO_x. Environmental Emissions Systems, Inc., 1992.

Index